IMPACT OF TECHNOLOGY ON CROP PRODUCTIVITY

IMPACT OF TECHNOLOGY ON CROP PRODUCTIVITY

DR. BALASAHEB PAWAR

PUBLICATIONS

KALPAZ PUBLICATIONS

DELHI-110052

Impact of Technology on Crop Productivity

Rs. 720/-

ISBN: 81-7835-519-1

M. Sc. (Agri.), Ph. D., M. A. (Econ.)
Department of Agricultural Economics,
Mahatama Phule Krishi Vidyapeeth, Rahuri-413722

Published in 2007 in India by
Kalpaz Publications
C-30, Satyawati Nagar,
Delhi-110052
E-mail: kalpaz@hotmail.com

Lasser Typesetting by: Rudra Computer Graphics, Delhi
Printed at : Salasar Imaging System, Delhi

Contents

List of Tables

List of Map, Graphs and Charts

Charts and Graphs

Prologue

The last decade of 20th century witnessed developments relating to liberalization of economic policy and formation of WTO. It is believed that once implementation phase (1995-2004) is over, countries having comparative advantage in producing a commodity will dominate the trade in that commodity. Fortunately, India being endowed with varied climates and soils, produces diversified crop-range. New technologies supported by appropriate services and public policies have led to the agricultural revolution (the green revolution), becoming one of the most significant of the scientific and socially meaningful revolutions. Indian agriculture is passing through an era of modern and ultra-modern technologies. With the optimum combination of crop enterprises and judicious allocation of allocation of available resources, there exists scope for increasing agricultural production.

The pivotal role of agricultural technology for increasing crop productivity in areas having uncertain rainfall needs no emphasis. In the present monograph 'Impact of Technology on Productivity of Crops in Maharashtra', authored by Dr. B. N. Pawar, an attempt has been made to investigate the level of adoption of technology, resources use pattern and allocation efficiencies in the different agro-climatic zones of Maharashtra, during pre-technological (1980) and post-technological period (2000), considering a edging line of implementation of NARP. He has succeeded in measuring the influence of agricultural technology on productivity of important crops in the State.

This book will prove useful for designing suitable production strategies for boosting the productivity of the crops. It will also be of use to the government for framing the price policies and for planning and implementing different development programmes.

R. B. Deshmukh

Preface

This book analyses the access of the relative aspects of the adoption level of technology, per hectare resource use, costs and returns, resource productivities, allocation efficiencies, and optimum use levels, share of technology in crop productivity and the economic sustainability of the technology according to different agro-climatic zones of the State. It is an approved Ph.D. thesis.

Author provides a detailed overview of technological impact on crop productivity during pre and post technological period. To what extent the technology contributes its share in the increased productivity and whether the available technology is sustainable. These aspects are critically evaluated in this book with the help of econometric models, various facts and figures and cross sectional data available (1980-81 and 2000-01). The author concludes that the magnitude of the technology adoption index was less than 50 per cent. There exist the disparities in the spread of modem technology in agriculture among the different agro-climatic zones. The decrease in the production elasticities in human labour, bullock power and seed showed the relative importance of process of transformation of agriculture from traditional to modem methods, whereas the increase in the production elasticities of manure and fertilizers, other working capital and technology adoption index indicated that there is a greater scope for intensifying applications of these inputs in agricultural production in the zones. The total change in per hectare crop output is the effect of crop production technology. The sustainability index brought out that the magnitude of yield realized by crop production method adopted during the year 2000-01, is sustainable. However, the full dissemination of the agricultural technology is not observed across the agro-climatic zones.

The author suggested that the efforts should be made for evolution of short duration, high fertilizer responsive, disease resistant, high yielding varieties of different crops. The research efforts aiming at evolution of low cost technologies of foodgrain crops suitable for rainfed conditions are expected to contribute to increased production of foodgrains. The efforts to foster increased use of plant nutrients and plant protection measures would help the grower to increase farm productivity. Immediate efforts are needed for evolving a breakthrough in the production technology of the crops for which, productivities have become stagnant and for optimizing the use of different inputs to increase the productivity.

Agriculture in developing economies was believed to be tradition-ridden and hence non-responsive to economic stimuli. The belief is that the response of agricultural production to changes in price was absent or perverse dominated for a fairly long time. In the context of developing economy, rapid growth of agricultural production assumes critical importance due to rising demand for them, generated by the growing population and rising levels of income. To meet the accelerated demand for agricultural output, the potential of agricultural sector has to be expanded and the flow of resources into agriculture has to be increased.

This book is presented in seven chapters. The first chapter is devoted to the nature of Indian agriculture, problems of agriculture in Maharashtra, importance of technology, the specific objectives, hypotheses and scope of the study. The second chapter deals with the methodology adopted. It includes selection of sample and the nature of the data required for the study and gives the methodological framework followed, which includes tabular analysis, production function analysis and economic analysis. The geographical features and agricultural economy of the Maharashtra is depicted in chapter three. The fourth chapter deals with resource use structure, production costs and returns, impact of agricultural technology on productivity of major crops in Maharashtra. The economic analysis pertaining to the resource productivities and allocation efficiencies in the production of major crops are presented in chapter five. It includes the production function model with economic optimal use levels of resources, shares of technology in crop production and sustainability of technology in crop production. The sixth chapter recapitulates the main findings of the study and elucidates the policy implications coming out of the present study. The last chapter presents the bibliography of related and important existing studies on the topic.

Acknowledgements

As a matter of fact, the study work was rather vast, tardy and difficult. I could complete this book mostly because of the co-operation, guidance and encouragement that I received from my colleagues, friends and relatives, to whom I am highly indebted.

I feel extremely honored for the opportunity bestowed upon me to carry out Ph.D. research under the guidance of Dr. Pandurang Mahadev Kapase. I wish to place my sincere heartfelt gratitude for his scholastic guidance.

My primordial obligation is to extend my gratitude to Dr. S. S. Kadam, Vice-Chancellor, Marathawada Krishi Vidyapeeth, Parbhani, Dr. R. B. Deshmukh, Vice-Chancellor, Mahatma Phule Krishi Vidyapeeth, Rahuri and Dr. Purushottam Bhapkar, (IAS), Chief Executive Officer, Zilha Parishad, Pune, Dr. Ankushrao Jadhav, Dean Faculty of Agriculture, MPKV, Rahuri, and Dr. J. R. Pawar, Ex-Head, Deptt. of Agricultural Economics, MPKV, Rahuri, for their encouragement at various stages of research work, valuable counsel, co-operation and suggestion for this manuscript.

I am also deeply obliged to all the authors, whose literature has been cited. During the course of this study, I have received direct as well as indirect help and assistance from a number of persons and organizations, for whom, I owe many thanks.

The enthusiastic, cheerful and selfless support of my dear friends through both the bright and dark phases of my work, has been invaluable to me. I offer my sincere thanks to Dr. Kripakar Wasnik.

A special mention must be made to Shri Amit Garg, Gyan Books Pvt. Ltd., New Delhi, who took the necessary pains for publishing this book in neat and elegant form within a very short period of time.

Last but not the least, words are small trophies in expressing my heart feelings to my wife Mrs. Rajashri and beloved son Ma. Ranjankumar and daughter Miss Pallavi, who always inspired me in all the facets of life.

Balasaheb Pawar

Date : 20/11/2006.

1

Introduction

The stagnation in agricultural production and levels of production of crops prevailed in the different regions for long period of early years of development. The way to stimulate agricultural growth in the developing countries is to raise investment opportunities by developing and supplying new input factors that embody improvements in technology and are therefore, profitable for farmers to adopt. (Schultz, 1964). The most practical and economical approach to achieve sizable increase in crop productivity and agricultural output therefore, lies in enhancing the efficiency of the existing agricultural economy through the introduction of modern technology on a broad front and on large scale.

1.1 Indian Agriculture

Agriculture is the heart and soul of India's socio-economic growth. Agriculture in India not only provides food for a population of more than one billion, but also provides livelihood to nearly 63 per cent of its population. Agricultural production in India has close association with total rainfall and its distribution. The crop production is very uncertain due to erratic and uncertain behaviour of rainfall. The occurrence of frequent and long dry spells, which results into unavailability of adequate soil moisture during the active growth period of crops, is the main reason for low and highly unstable yields. Traditional agricultural system apparently sustainable at low productivity and at low population pressure is breaking down under onslaught of high human and animal population pressures and unable to meet the changing demands of the society. A paradigm shift in soil and water management research and development is an inevitable necessity. This is possible provided that the uses of traditional knowledge and farmers' perceptions are woven around technology. The growing urbanization, industrialization and civic needs are creating new problems, pushing agriculture to more fragile environment and adding new dimensions to agenda of research and development in soil and water management.

In India, out of 328.7 million hectare geographical area, already 142.5 million hectare (about 47 %) is net cropped area, which is almost the upper limit to area extension for agriculture and there is no option except vertical expansion i.e. increasing productivity per unit area, per unit time by increasing the efficiency of essential farm inputs, be it fertilizers, irrigation water, energy, power etc. The supply of resources is not increasing in the required scale, while with the rise in population, the demand for employment is going up at a high rate. New employment opportunities need new resources for investment. Under these circumstances to feed the nation adequately, it is imperative to improve the production and productivity of crops on sustained basis, and one would hope that a proportionate resource use in agriculture would help to change existing situation by creating new rooms to accommodate the surplus rural population.

India being predominantly agricultural country wherein an economic development largely brings an increase in agricultural production. With the rapid increase in population, the demand for agricultural products increases speedily. The unabated population pressure is negating the effect of green revolution of late 'sixties' and 'seventies'. With the limited scope for expansion of land area under cultivation, significant increase in agriculture is possible only by raising productivity. The disparity in productivity of different agro-ecologies, declining growth rate in the green revolution areas, declining water table in tube-well areas, increasing soil degradation and growing apprehension about environmental safety with high input-based technology are the matter of concern for productive and sustainable agriculture. Efforts made in the absence of technological break-through in agriculture, the various institutional reforms practiced during 'fifties' proved to be sterile and ineffective in the attempt to increase agricultural production (Dantwala, 1970). As a result, the country was facing the problem of shortage of agricultural production, particularly of foodgrains and had to depend on imports in order to fulfill internal demand. Judging from the past experience and the experience of the countries, which have achieved high productivity growth rate with input-based technology, it is confirmed that without the use of input-based technology and improved management practices, India cannot move from the traditional low productivity system to continuously increasing productivity and sustainability, and this should be achieved without detriment to quality of environment especially of soil and water (Kanwar, 2000).

The various developmental activities in developing farming technologies of late 'sixties' and the 'seventies' brought about technological change in agriculture through adoption of improved production inputs. This increased agricultural production has enabled India not only to achieve self-sufficiency in foodgrains but also to possess adequate buffer stock. The achievements of green revolution brought about breakthrough in foodgrains production in India, it was confirmed mostly to some parts of the country that too in wheat and paddy crops. Moreover, this Green Revolution did not help to increase production and productivity of all the crops throughout the country.

Generally, the boons are not given free, warnings have already been voiced as to the possible social and political repercussions that could follow if new technological change is allowed to run its own course. Increasing rural unrest has been related to the rapid introduction of modern technology. The problem created by recent technological change can be categorised into three groups. The first-generation problems relate to physical constraints on the wide spread of new varieties imposed by weather conditions and state of advance in irrigation facilities associated with the pest control technologies, entrepreneurial skills of the farmers and on the levels of the input supply. The second-generation problems relate to the whole issue of marketing infrastructure, the re-organization and dynamism, which the technological change has necessitated. The third-generation problems assume added importance in the Indian context due to high population growth with already higher population density, very average income levels coupled with the great regional and personnel disparities in income and wealth (Falcon, 1970). Till recently, the official thinking was that the only feasible as well as surest way of improving the economic conditions of the farming community is to encourage the faster agricultural growth through subsidization of chemical and biological breakthrough, high yielding crop varieties, and intensive use of fertilizer and insecticides in production and through the promotion of agrarian capitalism into the country side.

New technologies supported by appropriate services and public policies have helped to prove doomsday prediction wrong and have led to the agricultural revolution (the green revolution) becoming one of the most significant of the scientific and socially meaningful revolutions of this country. Four thousand years of wheat cultivation tend to Indian farmers producing 6 million tonnes of wheat in 1947. In 1999-2000 our farmers harvested wheat over 72 million tonnes. This illustrated the power of technological change. In several crops, especially wheat, Indian farmers have made striking progress, taking the country to second position in the world in wheat production. The position in pulses illumines the pathway for a new strategy in agriculture. India occupied the first position in the world in both area and production of pulses, but 118th position in productivity. A major reason for the low average yield is the cultivation of pulses mostly under conditions where, soils are both, thirsty and hungry (Swaminathan, 2000).

The foodgrain production has touched 212.02 million tonnes in the year 2001-02 from the base of 50 millions tonnes in the early fifties. It is estimated that the total foodgrain requirement will be 243 million tonnes at the end of 2006-07 considering the India's population growth of 2.92 per cent per year. India has just crossed the population of one billion mark by the turn of this century and will be stabilized around 1.5 billion by the year 2030. This implies that the per capita cultivable land holding was 0.14 ha. in 2000 AD and it will be about 0.10 ha. in 2020 AD (Anonymous, 2004).

In short, the fruits of the green revolution have not been reaped uniformly in all the major crops, in all the parts of the country. This was because of the differentials in the climatic factors and regional inequalities of natural resource endowments across different regions in the country. Agriculture is characterized by wide spatio-temporal variation in the productivity of crops due to uncertainty and high fluctuations in climatic conditions. To bridge the gap between actual and potential yield prevailing at currently available level of technology, we have to undertake a multi-disciplinary analysis in different regions and farming systems. The regional differentials in production and productivities of different crops fetched attention of the policy makers to overcome the regional imbalance in agriculture. The ICAR, New Delhi launched National Agricultural Research Project during 1984-85 with an object to strengthen the research capabilities of NARS in the country by providing manpower and infrastructure facilities. In the country, total 131 agro-climatic zones were identified and strengthened so as to develop location specific technologies. Under this programme, emphasis was given to strengthen Research-Extension Linkage so as to accelerate the process of diffusion of technology. This has helped for pushing production and productivities in different agro-climatic zones of the country.

1.2 Problems of Agriculture in Maharashtra

Maharashtra is the third largest State in area and population, and forms major part of peninsular India with nine agro-climatic zones. The state has been divided into four socio-economic regions namely, Konkan, Western Maharashtra, Marathwada and Vidarbha. The variability in topography, soils and climatic factors bear significant impact on cropping pattern and technological innovations in crop production among the agro-climatic zones. In effect, there existed great disequilibrium in the bio-chemical and mechanical innovations among different agro-climatic zones as well as among individual crops within the zones.

The Maharashtra has wide agro-climatic zones based on varying rainfall pattern, topography, soil characteristics, climate and cropping pattern. The nine agro-climatic zones have been identified in the state. The technological advances in agriculture have not made much headway in the contribution to increased production and productivity. The State has all the while, remained on the forefront in adopting various developmental strategies including institutional reforms in order to increase agricultural production. However, the contribution made by these strategies has been observed to be relatively more conspicuous in the limited areas, where cultivation of sugarcane and cotton crops is predominant. The major areas of Maharashtra, however, could not make much progress in modernizing agriculture because of their excessive dependence on natural (agro-climatic) factors. The crux of the problem of modernization of agriculture in these areas is observed to be associated with the availability of capital resources, efficiency in allocating these limited resources to different uses, nature of technology and extent of its adoption determined by agro-climatic considerations on different types of farms.

The significant changes in biological and economical environment consequent upon the advent of improved technology in agriculture in Maharashtra have provided new alternatives, which in turn, have brought up the need for various adjustments in allocating limited funds among the competing resources. Many recent studies have shown that there were serious imperfections in resource allocation patterns of the farms as a result of mal-adjustment and lack of technical know-how.

Kapase (1992) stated that the per hectare use levels of crucial inputs viz; human labour, plant nutrients, quality seeds, etc. was observed to be higher in the zones having assured and high rainfall. There was also great level of inefficiencies in the use of different resources in the production of individual crops in different agro-climatic zones. The agro-climatic differences among the zones in the State led to wide variations in the per hectare resource use structure, costs and returns of individual crops. There is a need for evolution of suitable low cost production technologies for individual crops specific to different farming situations in the zones.

1.3 Impact of Technological Change in Maharashtra

There are however, some doubts regarding transfer and benefits of new technology over the farms and among the different agro-climatic zones. The new technology has potential to raise output but there are number of constraints viz., technical, economic, social as well risk and uncertainty involved in shifting from traditional to modern inputs, which have come into the way of wider adoption of new technology. Besides, highly variable agro-climatic factors have also caused considerable diversity in the impact of technological change in different parts of Maharashtra. Reliance on new agricultural technology has admittedly raised aggregate output, but at the same time it has benefitted only the large farmers who possess the requisite resources to use the new technology. On the other hand, resources restraints, particularly in regard to credit and working capital, and the risk and uncertainties involved in shifting from traditional to modern inputs have came in the way of wider adoption of the new technology by the small farmers (Sen, 1972). The areas with assured supply of water have benefitted more by new technology than by dry-farming areas. Whereas, high yielding varieties have been developed for irrigated areas, suitable varieties are still not available, which can give high yields under rainfed conditions (Pawar, 1978).

On the whole, the distribution of benefits among the agro-climatic zones would likely parallel existing resource endowments. In most of the agro-climatic zones, majority of the farmers are still lagging behind not only in full exploitation of this opportunity but also in proper use of their limited resources. Reasons for such situations can only be given by generating the knowledge of resource use structure, productivities and allocation efficiency of the farms in different agro-climatic conditions.

1.4 Importance of the Topic

In the 21st century, technological innovations in agriculture are going to be the most important determinants of the level, pace and sustainability of agricultural development. India has an impressive infrastructure for agricultural research comprising of 45 ICAR research institutes, 30 National Research Centres, 9 Project Directorates, 4 Bureaus, 28 SAUs, and one central agricultural university. They together employ more than 30,000 scientists, perhaps one of largest pools of scientists engaged in agricultural research in the world (Paroda, 1998). We have fairly good stock of new technologies available within the country but the level of adoption is low and skewed. Besides, there has been indiscriminate use of some of the new technologies such as chemical fertilizers and pesticides. The general economic and political environment prevailing in the country tends to favour and promote environmentally harmful technologies. It is therefore, necessary for ensuring safe and healthy lifestyles that environmental impacts of new technologies are carefully evaluated before they are recommended for wider use. The developing countries need to generate their own technologies suited to their conditions. The process of generating appropriate technologies, upgrading traditional ones, selecting and adopting improved technologies should take into account socio-economic and environmental factors (Singh, 1998).

The "Green Revolution" is the most striking success of development schemes of the post-independence era. The green revolution experienced in India in the last three decades and the achievement of self-sufficiency in foodgrains represent a success story of new agricultural technology in the country. After the green revolution years, it was realized that unless the differing farming situations based on micro-analysis are identified, diagnosed and remedied, the benefits of research programmes to improve the overall economy of farming community and thereby the nation would be difficult to ensure. The technological improvement depends upon the nature of underlying physical resources. The other factors like different types of soils, topographical situations, rainfall and its distribution, other climatic factors, socio-economic conditions of the farmers, social factors etc. also influences the level of technological improvement in crop production. The farming systems vary from one region to another depending upon variations in agro-climatic zones. Besides, limitations on the environmental control prevent farmers from achieving the much higher yield potential embodied in the new varieties of different crops. The regional variations are brought about due to variations in the agro-climatic conditions and differentials in the past investment in developing infrastructure. Hence, the National Agricultural Research Projects were launched to generate location specific, need based, production-oriented technology suitable to corresponding zones.

Technological change generates indirect employment through the backward and forward linkages in output and input markets in addition to the labour employment on farms. Technological change may also affect the relative economic position of the factors of production in terms of their share in total output and this implication for personal income distribution.

The technological change gives rise to different dimensions of income distribution problem such as regional, farm size groups and functional. The economic evaluation of effects of technological change on crop productivity and its implication helps in designing fiscal measures to mobilize part of increased income for meeting the goals of equal income distribution. The study on resource use efficiency and technological change assumes crucial importance for attaining growth in production of crops.

ICAR Institutes and State Agricultural Universities (SAUs) and many other government, semi-government and private organizations are involved in continuous economic well-being of the farming community. The SAUs play a vital role in the development of technologies appropriate for region, climate and needs of the farmers. These research institutions spend huge amounts on research and extension and evolve number of technologies. These technologies evolved after a series of scientific experiments and scrutiny at various scientific forums are propagated as recommendations by various research agencies for the adoption on the farmer's fields.

Some location specific technological practices have been identified and suggested to the farmers. But the productivity of crops has not much improved. There is a wide gap in the yield of crops per hectare between trials in research station (Haffis et al., 1990). The location specific technological advance in agriculture could not make much headway in their contribution in increased production and productivities in different zones. The major area of Maharashtra spread over these zones could not make much progress in modernizing agriculture because of their excessive dependence on natural (agro-climatological) factors. The benefits of the location specific technology and, strong research and extension linkage are noticed in various parts of the state, which has been pointed out by the research workers. Similarly, the technologies developed under various agro-climatic zones of the state must have benefitted the farming community in increasing the crop productivity and income.

Economic development is unsustainable if it increases the vulnerability of people to crises. A drought may force farmers to sell their animals or other productive assets needed for production in future years. A pestilence may damage a crop wholly and leave the farmers bankrupt. Similarly, a drop in prices of farm produce may reduce farm incomes and thereby cause farmers to over-exploit natural resources. In view of this, new technologies are needed to reduce the dependence of agriculture on climatic factors and thereby insure farmers against weather-induced risk and uncertainty. The farmers and policy makers still have practical needs and face problems that need practical solutions. The time has came that agricultural economists need to delve deeply into the assessment of the existing agricultural technology and find out the most economically viable and feasible technology which can withstand the Indian farmer in the globally liberalized world market. Productivity improvement will be possible only if we pay greater attention for improving the efficiency of input use, particularly nutrients and water.

The technology of biological products is mostly dynamic one as it changes with passage of time owing to release of new technology the improving earlier one, or removing its constraints, etc. There is need to carry out evaluation studies for assessing the impact, spread and extent of technology adoption from time to time besides, the need for identification of outdated technologies and searching for the new areas of research leading to betterment of the farmers. In the State, meagre efforts are made to study the impact of technology on crop productivities, employment, income etc. in case of some crops and in some parts of the states. Such studies have shown that technological improvement and linkage between Research and Extension certainly benefitted the farming community from the point of view of increasing the productivity, resource use efficiency, allocation efficiency, cost and return structure etc. In the state limited attempts have been made in this direction after implementation of NARP programme.

In view of the above, it was felt necessary to select the present topic for the dissertation. An attempt was made to examine the technological impact on productivity of major crops and the pattern of resources use structure, resource produtivities, and allocation efficiencies in the different agro-climatic zones of Maharashtra. The present study entitled **Impact of Technology on Productivity of Major Crops in Maharashtra** was taken up with the following objectives, in order to understand new production functions whose primary aim is to increase resource use within a given technological environment.

1.5 Objectives

In the light of the above observations the main objectives of this study are stated as under.

(i) To assess the level of adoption of crop production technology in different agro-climatic zones.

(ii) To find out technological impact on use of crucial inputs and productivity of major crops in different agro-climatic zones.

(iii) To study the changes in costs and returns structure of major crops over a period of time.

(iv) To examine the changes in resource use efficiency and determine the optimum use levels in production of major crops.

(v) To find out the share of the technology in the productivity of major crops over a period of time.

(vi) To assess the economic sustainability of crop production technology over a period of time.

1.6 Hypotheses

In the sphere of the foregoing theoretical propositions and the review of the literature presented in the next chapter, the following null hypotheses were proposed to be tested for fulfilling the requirements of the objectives of this study.

1. Technology adoption level was increased in all the agro-climatic zones over a period of time.
2. Because of technological impact, there was an increase in productivity of major crops in different agro-climatic zones.
3. There was a positive change in the costs and returns of major crops over a period of time.
4. There existed changes in resource productivities and resource use efficiencies of major crops over a period of time.
5. The increased crop productivity was associated with the impact of technologies during the period under study.
6. The crop production technologies were sustainable and could improve the economic benefits on the farmer's field.

1.7 Scope and Utility of the Study

Today, Indian agriculture is certainly passing through era of modern technology. Modernization of traditional agriculture involves the intensive use of improved technology, the development of farm resources, and infrastructures within and outside the farm. The rapid technological advances envisage higher level of capital requirements in managing the farm production. The use of this new strategy implies the application of heavy doses of fertilizers, intensive irrigation and the use of high yielding varieties of seeds, which essentially mean higher capital requirements. But, the own resources of the farmers are generally insufficient to meet the rapidly growing capital needs. Therefore, a sustainable farm planning with rational decisions concerning to the best combination of enterprises is inevitable because of great variation that exists in the quantity and quality of resources available under different farm situations in different agro-climatic zones, which directly contribute to the productivity. The increase in agricultural productivity is essentially a result of management decisions made by the individuals. There is thus ample scope to increase farm income by allocating the farm resources at optimum level. The modern agriculture, a high level efficiency with regards to optimum combination of crop enterprises and allocation of available resources, has great significance for increasing agricultural production. This situation calls for immediate and urgent attention of experts in the field of agriculture so as to bring about optimum allocation of resources to the extent possible.

Agricultural development in India has initiated the process of shift from traditional to modern farming systems. If the pattern of resource use is inefficient, production can be increased through reallocation of resources in the optimal direction. In case, it is efficient, then only way for increasing production will be breakthrough in technological change. The adoption of new agricultural technology is a production syndrome encompassing the use of high yielding varieties, chemical fertilizers, insecticides and non-traditional farm management techniques. The characteristic of modern technology in agriculture is the application of more capital inputs consequent to technological changes such as, new varieties of seeds, fertilizers, pesticides, machineries, etc. that helps to raise the productivity of resources. It also enhances the scope to substitute inputs of elastic supply such as fertilizers with those having inelastic supply such as land.

As indicated earlier, due to variations in the agro-climatic conditions and differentials in the past investment in developing infrastructure, the National Agricultural Research Projects (NARP) were implemented to minimise the problems of farming community in different agro-climatic zones. The efforts were found to be successful since at micro-level impacts have been noticed. However, the need of micro-level studies was felt necessary to quantify the impact of technology. The purpose of this investigation is to study the economic aspects viz. resource use productivity, cost structure and profitability of major crops in different agro-climatic zones of Maharashtra. These will perhaps, serve as feedback points to the scientists for modifying the technology in order to make it more adaptable by the farmers. The findings of the study would be useful to the cultivators in knowing present crop production technology and possibilities of increasing income through reallocation of resources. This will be useful for designing suitable production technologies for boosting the productivity of the crop. The study will be important from the point of view of theory as well as practical side of the agricultural production field.

This study will serve as guidelines for utilizing the optimum level of available resources for production of major crops. The study will be useful to the financing agencies for advancing loans to the cultivators. It will also be of great help to the government for framing the price policies and, for planning and implementing different development programmes.

In addition, it is also an attempt to illustrate the use of production function analysis in crop enterprises. The results of the productivity of different resources in different regions may be used by the policy-makers to assess the possibilities of intensifying the resources in areas and classes of farmers where, the specific resource commands the productivity. The knowledge of the productivities of resources may also be used since it provides guidelines for adopting suitable development strategies.

1.8 Limitations of Study

In view of the limitation of the time and manpower available at the disposal of the author, the investigation was restricted for 397 sample farmers each for two points of time, i.e. pre-technological and post-technological period of implementation of NARP programmes in the State. The sample was drawn from nine agro-climatic zones spread over entire Maharashtra state. The selected sample farms were rainfed which, mainly represented the vast exposes of the chronically affected scarcity zone of the State. It does not take into account the conditions in the relatively more affluent parts from the command areas of irrigated agriculture in the State. Since, the study was based on the limited number of the farmers, from each of the agro-climatic zones, the results of the study may not be undertaken as representative of all parts of the country, however, it would be indicative in the parts of the country with similar situations.

In this study, an attempt was made to quantify the impact of technologies released under NARP approach by the State Agricultural Universities. This development is in its initial phase. There is large scope for improving this methodology. The determination of weights, approach for development of composite index, factor shares, sustainability analysis and year to year variations in weather and biological outputs need to be explored.

1.9 Presentation of Study

This study is presented in six chapters. The first chapter is devoted to the nature of Indian agriculture, problems of agriculture in Maharashtra, importance of technology, the specific objectives, hypotheses, scope of the study and limitations of the study. The second chapter presents the brief reviews of related and important existing studies on the problem according to the sub heads. The third chapter deals with the methodology adopted. It includes selection of sample and the nature of the data required for the study and gives the methodological framework followed, which includes tabular analysis, production function analysis and economic analysis. The geographical features and agricultural economy of the Maharashtra is depicted in chapter four. The fifth chapter deals with resource use structure, production costs and returns, impact of agricultural technology on productivity of major crops in Maharashtra. The economic analysis pertaining to the resource productivities and allocation efficiencies in the production of major crops are presented in chapter six. It includes the production function model with economic optimal use levels of resources, shares of technology in crop production and sustainability of technology in crop production. The last chapter recapitulates the main findings of the study and elucidates the policy implications coming out of the present study.

2

Research Methodology

Every research project has to be planned and adopted with appropriate methods and procedures to obtain the desired results of the scientific study chosen. This chapter deals with the explanation of methods and techniques used in conducting the present study. The major topics dealt in this chapter are research design and locale of the study, nature of respondents, procedure used for collecting the data, development and use of measures for general level concepts and variables alongwith the operational definitions of terms used. The last part deals with the details of the statistical tools used for the compilation and analysis of data. The methodology adopted for the present investigation is narrated below.

2.1 The Data Requirement

The implementation of NARP during the year 1983-84 to 1996-97 is the yardstick for pre-technological and post–technological period. The year 1980-81 and 2000-01 (being the normal years) were treated as pre-technological and post-technological period, respectively. As the study was related to examine the impact of technology on productivity of major crops in different agro-climatic zones, the data for two points of time were needed. The data pertaining to resource use, cost and return structure, resource productivities and allocation efficiency in agriculture, for two points of time, i.e. for the year 1980-81 and 2000-01 were contemplated. This time-lag of twenty years was considered to be sufficient to examine the impact of technology on productivity of crops and efficiency of allocating resource on the part of the farmers in the region. As such, the requirement of data was with respect to structure of the sample farms, physical inputs and outputs of the crops, prices of inputs and outputs and selected indicators of the State agricultural economy.

2.2 Locale and Research Design

Inadequacy of data in India is a big handicap for any economic research work. It is more so particularly in the sphere of farm business research. Fortunately, the Government of Maharashtra, recognized the dearth of adequate information pertaining to farming business in the State and started the 'Scheme for Creating Permanent Machinery for studying the Cost of Cultivation of Principal Crops in Maharashtra' in the year 1979. This scheme is proved to be successful in providing the sufficient database for farm business analysis and formulation of appropriate agricultural price policies. The present investigation was conformed to the database of the said scheme. The sample farms in different clusters of villages spread over in agro-climatic zones of the State, selected under Government of Maharashtra sponsored Cost of Cultivation Scheme, were taken up for the present study.

As such, the sampling design adopted in the scheme consisted of three-stage stratified random sampling plan with homogeneous crop growing zones as strata, tahsil as a primary unit, a cluster of villages as secondary unit and land holding within the cluster as an ultimate unit of sampling. The above scheme is operated by the Department of Agricultural Economics in the four Agricultural Universities in the State. A huge amount of data relating to the cost and returns of selected crops are being collected by the cost accounting method, which ensures the accuracy and reliability. The study covers quite a substantial number of holdings selected by adopting a scientific sampling technique, keeping in view of maintaining the adequate sample size. The crop complex as well as individual crop approach was adopted in collection of data under the scheme.

2.3 Selection of Villages and Sample Cultivators

The exhaustive list of the cluster villages selected under the said scheme for the year 1980-81 and 2000-01 was made according to the agro-climatic zones. During the base year (i.e. 1980-81), eight crops were studied under the above said scheme and *suru* sugarcane as well as *ratoon* sugarcane was studied under the Central Government sponsored 'Comprehensive Scheme for Cost of Cultivation of Major Crops' as this crop was not covered under the State sponsored Cost of Cultivation Scheme during the year 1980-81. Based on manned clusters, the cluster villages and cultivators for respective crop were selected for both the period. In all for nine crops, 397 sample farms were selected for each of the period i.e. pre-technological and post-technological period. The precaution was taken that agro-ecological and socio-economical features remain the similar for sample farms selected in 2000-01 to avoid the differences in the socio-economic conditions within the sample farms selected during the base year (1980-81).

The distribution of the sample farms according to the agro-climatic zones indicating the districts, tahsils, villages and number of holdings are indicated in Table 2.1. The cropwise as well as agro-climatic zonewise distribution of total sample farms is shown in Table 2.2. The location of the tahsils in different agro-climatic zones of Maharashtra is shown in between the page No 40 and 44 on the map to indicate the geographical distribution.

2.4 Collection of Data

The farm level data collected by cost accounting method under the above said schemes, were scanned for two points of time i.e. for the years 1980-81 and 2000-01, being the normal years from the climatic point of view. The data of the former year were treated as base line data to estimate the impact of agricultural technologies on productivity of major crops.

The crops, viz; *kharif* sorghum (unirri.), *kharif* paddy (unirri.), pearlmillet (unirri.), groundnut (unirri.), cotton (unirri.), *rabi* sorghum (unirri.) and wheat (irri.), were studied. Sugarcane crop, which plays a significant role in the economy of the Maharashtra, was also taken up for the study. For *suru* sugarcane and *ratoon* sugarcane crop, the data for the year 1980-81 were scanned from Comprehensive Scheme for Cost of Cultivation of Principal Crops in Maharashtra sponsored by Government of India and operated by Department of Agricultural Economics, MPKV, Rahuri. The main focus of the study was on the agricultural features of sample farms in different agro-climatic zones and their structure. The detailed information on the aspects, such as farm family organization, land holding, cropping pattern, livestock position, farm implements and machinery tools, labour use, use of different inputs, crop production, yield and value of main and by produce, technology adopted etc. were obtained from the primary schedules generated in the above-said schemes. The data on agricultural economy of Maharshtra were obtained from the published sources such as Epitome of Agriculture in Maharashtra, and Agricultural Statistical Information of Maharashtra.

Table 2.1 : Distribution of sample farms according to different agro-climatic zones

Crop	1980-81					2000-01				
	Zone	District	Tahsil	Village	Sample	Zone	District	Tahsil	Village	Sample
Kh. Sorghum (unirr.)	VII	Jalgaon	Bhusaval	Dhanora	7	VII	Jalgaon	Dharangaon	Dharangaon	6
		Jalgaon	Bhusaval	Khedgaon	9		Jalgaon	Parola	Tamaswadi	9
		Jalgaon	Erandol	Kalyane Bk	6		Jalgaon	Jalgaon	Nashirabad	7
					22					**22**
	VIII	Nanded	Hadgaon	Khirgaon	8	VIII	Yeotmal	Kalam	Sawargaon	7
		Yeotmal	Yeotmal	Bhari	5		Nanded	Hadgaon	Sawargaon	7
		Nagpur	Katol	Chorkhain	8		Nagpur	Katol	Zilpa	7
					21					**21**
Kh. paddy (unirri.)	II	Raigad	Mangaon	Gowel	10	II	Raigad	Karjat	Kashele	10
		Nasik	Igatpuri	Korapgaon	10		Kolhapur	Kagal	Murgud	10
					20					**20**
	V	Nasik	Dindori	Chachadgaon	10	V	Nasik	Dindori	Warkhed	10
		Satara	Karad	Malkhed	10	IV	Satara	Patan	Chafere	10
					20					**20**
	IX	Bhandara	Sakuli	Nimgaon	10	IX	Bhandara	Sakuli	Lakhani	10
		Chandrapur	Chandrapur	Gowardhan	10		Chandrapur	Chimur	Bhishi	10
					20					**20**

(Contd.)

(Contd.Table)

Crop	1980-81					2000-01				
	Zone	*District*	*Tahsil*	*Village*	*Sample*	*Zone*	*District*	*Tahsil*	*Village*	*Sample*
Pearlmillet (unirr.)	V	Ahmednagar	Akole	Pimpal-kunta	10	V	Satara	Satara	Upsinge	10
		Ahmednagar	Akole	Nimbral	10		Pune	Junnar	Khodad	10
					20					**20**
	VI	Nasik	Niphad	Nandgaon	10	VI	Nasik	Sinnar	Nandur-shingote	8
		Nasik	Sinnar	Musalgaon	10		Nasik	Nandgaon	Nyayadon-gari	12
					20					**20**
	VII	Jalgaon	Parola	Sangavi	10	VII	Jalgaon	Parola	Tamaswadi	6
		Jalgaon	Chalisgaon	Wadale-wadali	4		Jalgaon	Raver	Khiroda	8
					14					**14**
Groundnut (unirri.)	V	Kolhapur	Hatkanan gale	Sawarde	10	V	Kolhapur	Hatkanangale	Herale	9
		Pune	Junnar	Kuran	10		Pune	Junnar	Khodad	11
					20					**20**
	VII	Jalgaon	Parola	Morphal	10	VII	Jalgaon	Parola	Tamaswadi	10
		Beed	Ambejogai	Dainapur	10		Beed	Ambejogai	Badipur	10
					20					**20**
Cotton (unirri.)	VII	Jalgaon	Parola	Kol-pimpri	10	VII	Jalgaon	Parola	Tamaswadi	10
		Akola	Akot	Varud	9		Akola	Akot	Palsod	9
					19					**19**
	VIII	Parbhani	Parbhani	Makhani	10	VIII	Parbhani	Parbhani	Zari	10
		Yeotmal	Wani	Majara	10		Yeotmal	Wani	Mohada	10
					20					**20**

(Contd)

(Contd)

Crop	1980-81					2000-01				
	Zone	District	Tahsil	Village	Sample	Zone	District	Tahsil	Village	Sample
Rabi sorghum (unirri.)	VI	Dhule	Sindhkheda	Vasmane	10	VI	Solapur	Karmala	Jeur	10
		Solapur	Karmala	Alsunde	10		Nandurbar	Shahada	Sarangkheda	10
					20					**20**
	VII	Solapur	S. Solapur	Boramani	9	VII	Jalgaon	Jalgaon	Asoda	12
		Beed	Kej	Selu	7		Beed	Gevarai	Chakalamba Bk	4
					16					**16**
Wheat (irri.)	VI	Nasik	Yeola	Nevargaon	8	VI	Nasik	Yeola	Patoda	8
		Dhule	Sindhakheda	Karle	10		Nandurbar	Shahada	Sarangkheda	10
					18					**18**
	VII	Jalgaon	Erandol	Dharangaon	5	VII	Jalgaon	Jalgaon	Asoda	8
		Osmanabad	Osmanabad	Irala	13		Latur	Nilanga	Ambulaga Bk	10
					18					**18**
	VIII	Parbhani	Basmat	Kanosa	10	VIII	Hingoli	Basmat	Adgaon	10
		Nagpur	Katol	Hongandi	9		Nanded	biloli	Mugaon	9
					19					**19**
Sugarcane	VI	Ahmednagar	Newasa	Kharwandi	7	VI	Ahmed nagar	Newasa	Kukana	6
(Suru)		Dhule	Sakri	Sayane	5		Nasik	Niphad	Niphad	7
		Nasik	Niphad	Chandori	8		Dhule	Sakri	Pimpalner	7
					20					**20**
	VII	Solapur	N. Solapur	Darphal	5	VII	Solapur	Pandharpur	Gursale	6
		Aurangabad	Sillod	Mangrul	5		Auranga bad	Sillod	Andhari	5
		Osmanabad	Latur	Murud-Akola	6		Latur	Nilanga	Shirur-Anantpal	5
					16					**16**

(Contd)

(Contd)

Crop	*1980-81*					*2000-01*				
	Zone	*District*	*Tahsil*	*Village*	*Sample*	*Zone*	*District*	*Tahsil*	*Village*	*Sample*
Sugarcane	VI	Ahmednagar	Newasa	Kharwandi	7	VI	Ahmed nagar	Newasa	Kukana	7
(Rotoon)		Dhule	Sakri	Sayane	4		Dhule	Sakri	Pimpalner	6
		Satara	Phaltan	Nimblak	6		Satara	Koregaon	Waghli	4
					17					**17**
	VII	Auranabad	Sillod	Mangrul	9	VII	Auranga bad	Sillod	Andhari	9
		Osmanabad	Latur	Murud-Akola	8		Osmana bad	Tuljapur	Arli- Bk	8
					17					**17**
Total				**Total**	**397**				**Total**	**397**

Table 2.2 : Distribution of sample farms according to agro-climatic zones

Crop	*Year 1980-81*		*Year 2000-01*	
	Zone	*Sample*	*Zone*	*Sample*
Kharif sorghum (Unirr.)	VII	22	VII	22
	VIII	21	VIII	21
Kharif paddy (Unirri.)	II	20	II	20
	V	20	V	20
	IX	20	IX	20
Pealmillet (Unirr.)	V	20	V	20
	VII	20	VII	20
	VII	14	VII	14
Grundnut (Uniri)	V	20	V	20
	VII	20	VII	20
Cotton (Unirri.)	VII	19	VII	19
	VIII	20	VIII	20
Rabi sorghum (Unirri.)	VI	20	VI	20
	VII	16	VII	16
Wheat (Irri.)	VI	18	VI	18
	VII	18	VII	18
	VIII	19	VIII	19
Suru sugarcane	VI	20	VI	20
	VII	16	VII	16
Ratoon sugarcane	VI	17	VI	17
	VII	17	VII	17
Total samples	**All zones**	**397**	**All zones**	**397**

2.5 Analytical Procedure

This section deals with the methodology adopted for analysis of the data to arrive at meaningful results. The emphasis was given on the simple tabular analysis based on means, percentages, frequencies, etc.

2.5.1 Estimation of resource use levels

The requirements of major inputs and levels of output produced in respect of individual crops were worked out on per hectare basis separately for different agro-climatic zones. The comparison of per hectare resource use levels and output of crops for the year 1980-81 and 2000-01 could provide the basis for understanding the differentials in costs and returns and resource productivities of crops across the different agro-climatic zones at two points of time.

2.5.2 Estimation of production costs and returns of crops

The simple method of tabular analysis based on means and averages was followed in estimation of costs and returns of individual crops per unit of land for both the period. The changes in cost of cultivation and return structures of major crops resulting from adoption of technology were examined by the percentage change over the base year i.e. 1980-81.

The present study was related to two time periods with a time lag of twenty years. During this period, considerable changes in prices of all the inputs as well as output of crops might have taken place. The study of this kind is always subjected to give misleading picture about the farm economy, if the effects of price changes are not taken care of. In order to overcome the problem of price changes, the costs of inputs and returns from the crops, for the year 1980-81 were inflated at the prices that prevailed in the year 2000-01. The other costs items, such as depreciation and repairs on farm implements and machinery, incidental charges, land taxes, interest on fixed capital, etc. were inflated with the inflation factor of wholesale price index during the year 2000-01.

The method of estimation and allocation of costs are discussed in brief separately under the topic, viz., Definitions and cost concepts used in study. Total input costs were estimated at three stages of costs i.e. cost-A, cost-B and cost-C on the basis of standard cost concepts.

2.5.3 Assessment of technological impact

As discussed in the first chapter the crop production technologies were developed by SAUs for each crop grown under different agro-climatic zones. The same were recommended for accelerating the productivities of the crops and thereby higher returns to the producers. Sufficient efforts were made to disseminate the crop production technologies by the extension personnel of Government of Maharashtra as well as Government of India by developing linkage between researchers and farmers under the various programmes. The NARP approach was one of the programmes, which has observed to be the most effective one to develop better linkage between scientists and farmers. As a result, farmers' awareness and adoption behaviour were improved, which was reflected through the various studies cited in Chapter 2. The studies also indicated that there was clear change in productivity and returns. The changes in the crop productivities and returns were due to adoption of improved crop production technologies, which was considered as impact of technology.

Scoring index for selected components of technology

Input		*Use levels*	*Points*
Variety	a.	Hybrid	3
	b.	High yielding	2
	c.	Local	1
Seed rate	a.	Recommended level	3
	b.	More than recommended level	2
	c.	Less than recommended level	1
FYM /Fertilizers	a.	Recommended level	6
	b.	Above recommended level	5
	c.	76 to 99 % of recommended level	4
	d.	51 to 75 % of recommended level	3
	e.	26 to 50 % of recommended level	2
	f.	Up to 25 % of recommended level	1
	g.	No adoption	0
Plant protection	a.	Plant protection measures undertaken	1
	b.	Plant protection measures not undertaken	0

The recommendations involved in crop production are of two categories. The recommendation of components of crop production technologies such as adoption / non-adoption of seed treatment or split application of either of NPK etc. are categorised as 'qualitative recommendations' and other recommendations such as application of 50:25:0 kg. NPK per hectare are categorised as 'quantitative recommendations'. In this study, for determining the impact of agricultural technologies at two points of time, five key components, namely, variety, seed rate, FYM, fertilizers and plant protection measures were considered. The extent of adoption of recommendation by the farmer with respect to each component of crop production technologies was quantified as follows and expressed in the form of indices and they were termed as 'Technology Adoption Indices' (TAI). Following pattern was employed to assign the scores for ascertaining the use levels of inputs due to technologies.

Technology Adoption Indices

The quantitative component of technology, the production indices are defined as:

The Technology Adoption score for the quantitative component of the production for the selected crops was computed by Technology Adoption Index (TAI) as per Kiresur et al (1996) as below.

$$TAI = (A_i / M_i) \times 100$$

where,

A_i = Average adoption score registered by farmer for i[th] crop

M_i = Maximum adoption score for i[th] crop.

Development of Composite Index

The composite TAI was developed to represent overall adoption of crop production technology. The composite index of adoption of the technology was estimated for each farmer based on the adoption index of the technology and factor share of respective technology. For this purpose, a weighted linear combination of individual component of technology was expressed in the form of indices explained above were considered.

The composite index was developed as follows:

$$CAI_i = \sum_{i=1}^{n} W_i \times AJ_i$$

where,

CAI_i = Composite Adoption Index.

W_i = Weight assigned to the ith technology.

AJ_i = Adoption index of ith technology at recommended level.

2.5.4 Production function analysis

It was felt that simple tabular analysis based on means, percentages, frequencies, etc. suffers from some major limitations in the sense that it can not precisely measure the contribution of a specific resource / factor to output in combination with other resources which are responsible for changes in the level of output under given set of production technology. The technological change is one of the major forces leading to changes in output. The analysis of technological change can be carried out most conveniently by the use of theory of production. The production function approach does not have the above limitation. The production function analysis gives the marginal value productivity of resources at their geometric mean levels, which is more meaningful in the context of economic development, where the aim is to get maximum output from available resources. The efficiency of resource use under different farm situations as well as under different technological environments can be judged with the production function analysis. Therefore, the production function analysis was used in the present study.

2.5.4.1 Production function form

The production function describes the transformation of set of inputs to output. Production function thus, reveals a technical and mathematical relationship describing the manner and extent to which a particular product depends upon the quantities of inputs used. The empirical evidence from previous studies suggested that the Cobb-Douglas production function of the following type has been widely used in agriculture for studying relationship between crop output and input variables.

$$Y = a X_1^{b1} X_2^{b2} \ldots\ldots\ldots X_n^{bn} e^u$$

The Cobb-Douglas production function provides a direct test on the existence of the rational production behaviour. This is because, it accounts for only the so-called second stage of neo-classical production function. Furthermore, Cobb-Douglas production function completely disregards the existence of the third stage of production, which is characterized by zero or negative marginal productivity. It is the most appropriate one as it permits diminishing, increasing or constant returns to scale without using as many degrees of freedom as other functional forms. In this functional form, 'Y' is dependent variable, '$X_{i's}$' are independent resource variables, 'a' is constant representing intercept of the production function and '$b_{i's}$' are the regression co-efficients of the respective resource variables. The regression co-efficients obtained from this function are also elasticities of production, which remain constant through relevant ranges of inputs and individually they indicate the percentage change in output for unit percentage in the concerned input. The sum of elasticities i.e. '$b_{i's}$' indicates the nature of returns to scale. When expressed in logarithmic terms, this function transforms into a linear form of the following type.

$$\text{Log } Y = \log a + b_1 \log X_1 + b_2 \log X_2 + \ldots\ldots + b_n \log X_n + u \log e$$

The scale line or the expansion path in the Cobb-Douglas function is a straight line passing through the origin and the equation of the scale line can be derived by equating the Marginal Rate of Technical Substitution (MRTS) between factors with their inverse price ratio. With the help of this function, the marginal value product was estimated and the resource use efficiency of different inputs was examined at two points of time.

2.5.4.2 Specification of variables

A great deal of caution needs to be exercised in the selection, classification and aggregation of input variables used in production process for studying resource productivity. Farm inputs have been classified and aggregated in different ways by the various research workers so as to fulfil the requirements of the study and to keep the number of variables to a manageable size.

For fitting the production function, six inputs, namely, human labour, bullock power, seed, manure and fertilizers, other working capital and technological adoption index were considered as important factors in the production of crop. The output of the crop was used as dependent variable. The equation fitted was of the following form.

$$Y = a X_1^{b1} X_2^{b2} X_3^{b3} X_4^{b4} X_5^{b5} X_6^{b6} e^u$$

where,

Y = Output of main produce (qtls./ha.)
X_1 = Human labour (mandays/ha.)
X_2 = Bullock power (pair-days/ha.)
X_3 = Seed (Kg./ha.)
X_4 = Manure and fertilizer's cost (Rs./ha.)
X_5 = Other working capital (Rs./ha.)
X_6 = Technology adoption index (%)
u = Error term
a = Intercept
$b_{i's}$ = Regression coefficients

The significance of coefficients of each of the variables from the estimated function were tested with the help of 't' statistics and the overall significance of the fitted equation was judged on the basis of 'F' value.

The greatest difficulty with this type of analysis is that, if the factors themselves are highly correlated and no reliance can be placed on the regression co-efficient. In such cases, for any arbitrary regression co-efficient for one factor, a regression co-efficient of another factor can be determined, which will give desired results. This is called the multicollinarity.

From the correlation co-efficients, it is seen that all other factors are highly correlated with area under crop. Thus, if area is highly correlated with the other factors of production in any multiple regression, the multicollinarity will vitiate the result. To circumvent this serious shortcoming of the data, the non-area variables were converted on per hectare basis in the analysis. Another problem confronted in the production function analysis is the existence of non-significant variables. Some researchers have suggested that such non-significant variables be excluded from the production function analysis. However, some other researchers seem to be against this view. To put it in the words of Johnston (1972), exclusion of non-significant variables lowers the R^2 value considerably. This is because some regression coefficients capture the explanations provided by the omitted variables. It may also happen that for some other coefficients, the standard errors increase without a sizable increase in the coefficients themselves. The exclusion of non-significant variables may cause a bias in estimating the coefficients of the remaining variables. Also exclusion of non-significant variables would imply loss of, although imprecise, information. Besides, the practice of exclusion of non-significant variables from some production functions, and retaining the same in other production functions on the ground that they are significant (as is done in some studies), does not facilitate comparison among the fitted production function of the kind attempted in the present study.

A brief description of the inputs used as explanatory variables in the present study is indicated below.

(i) Human labour (X_1) : This input was expressed in terms of mandays of eight hours. It included all human labour (including operator's labour, hired labour, permanent and temporary, and other family labour) utilized for performing different farm operations right from preparation of land to threshing and winnowing. The differences in the efficiency of labour were taken in to account by converting female labour at the rate of one female day equivalent to 0.75 male day.

(ii) Bullock labour (X_2) : Bullock labour used for different farm operations was considered as a separate input and it was measured in pair-days. Hence, one pair-day means eight hours of work by the one pair of bullocks.

(iii) Seed (X_3) : Seed material used for sowing was measured in physical quantity (i.e. in kilogram).

(iv) Manure and fertilizers (X_4) : The input variable included the use of plant nutrients in the form of manure of organic and inorganic nature. The manure and fertilizers applied to the farm was measured in monetary terms.

(v) Other working capital (X_5): This variable was defined as expenditure on other inputs. The inputs like irrigation, insecticides, fungicides, minor repairs and maintenance of implements and machinery, etc were considered. All these inputs were aggregated into only one variable for the reason that each of these items individually may have a small association with production, and inclusive of all these items as separate variable would greatly reduce degree of freedom.

(vi) Technology adoption index (X_6) : Technology adoption score for individual crop was computed by the method as explained in earlier paragraphs and used in percentage as explanatory variable.

(vii) Output (Y) : The output was used as dependent variable. For individual crop, the output was expressed in physical quantities in quintals.

2.5.4.3. Estimation of Production Function

For estimation of resource productivity and determining optimal allocation of capital among the categories of inputs, a production function analysis was used. The production function analysis was attempted separately for each crop for the two time periods i.e. 1980-81 and 2000-01. The production function was estimated for 9 crop enterprises viz, *kharif* paddy (unirri.), pearlmillet (unirri.), *kharif* sorghum (unirri.), groundnut (unirri.), cotton (unirri.), wheat (irri.), *rabi* sorghum (unirri.), sugarcane (*suru*) and sugarcane (*ratoon*).

All the production functions were estimated using ordinary least square method with the help of computer programme. Also, the values of R (Coefficient of multiple correlation), R^2 (Coefficient of multiple determination), standard errors of regression coefficients and 'F' statistics (analysis of variance) were estimated for judging overall significance of the production function and regression coefficient as well.

2.5.5 Economic Analysis

The economic analysis of crop production activities can be carried out most conveniently by using theory of production functions. The estimated production functions were used for economic analysis with regard to estimation of returns to scale, marginal productivities of individual resource factors and optimal allocation of different resources on the farms. The differences in the value of these estimates gave an idea about the variations in the productvities of resources and allocation efficiency of the farmers for two points of time i.e. pre-and post-technological periods.

2.5.5.1 Returns to Scale

In the Cobb- Douglas production function, the returns to scale are measured by adding elasticities of all the inputs. Thus, for the production function:

$$Y = a X_1^{b1} X_2^{b2} X_3^{b3} \ldots . X_n^{bn}$$

where, $b_1, b_2, b_3, \ldots, b_n$ are the regression coefficients or elasticities of production and return to scale is,

$$\Sigma bi's = b_1 + b_2 + b_3 + \ldots + b_n$$

Depending upon the sum of regression coefficients, as less than, equal to or greater than unity, the prevailing returns to scale may be considered as decreasing, constant or increasing, respectively. Since the elasticity of coefficients are constant in the Cobb- Douglas production function, the scale of returns so indicated simply represent the average condition for the sample farms. The 'Z' test was used to test the significance of the difference of sum of production elasticities i.e. $\Sigma bi's$ from unity. The following formula was used to calculate 'Z*' value, (Koutsoyiannis, 1977).

$$\Sigma bi's - 1$$

$$Z^* = \Sigma bi's - 1 / \text{S. E. of sum of the production elasticities}$$

Here we wish to test the null hypotheses Ho : $\Sigma bi's = 1$, against the alternative hypothesis H1 : $\Sigma bi's$ '

If the empirical 'Z*' falls outside the critical regions at 5 per cent level of significance i.e. $-1.96 < Z^* < 1.96$, then we accept our null hypothesis,

Ho : $\Sigma bi's = 1$ and vice versa.

2.5.5.2 Estimation of Marginal Value Products (MVP)

In order to study the efficiencies of resource use for different crops, the marginal value product (marginal physical products in case of the crop enterprises) of the individual resource, was estimated and compared with its unit price (MC). The regression coefficients or elasticities of production of inputs thus, obtained were used to calculate the marginal value products of all resources at their geometric mean levels. The marginal value product of individual resource was estimated by using the following formula.

$$\text{MVP of } X_i = b_i \frac{\bar{X}_i}{\bar{Y}_i} Py$$

where,

b = Elasticity of production of ith input.
Y_i = Geometric mean of output
X_i = Geometric mean of ~~i^{th}~~ input
P_y = Per unit price of output.

The estimation of resource productivity among the categories of inputs, was studied through the comparison of MVP of resources with their respective acquisition costs. The significant difference between MVP of resources and their acquisition costs were tested by computing 't' values as follows:

$$t = \frac{\text{MVP } x_i - Px_i}{\text{S. E. of MVP } x_i}$$

where,

MVP = The marginal value product of ith resource.
Px_i = Acquisition cost of x_i th resource.

The S. E. of $MVPx_i$ is worked out by taking square root of $[(Y / X_i)^2 . V(b)]$
where, V(b) denotes variance of regression coefficient of X_i.

A production function analysis as mentioned earlier was used. The changes in efficiency of resource use under different crop situations resulting from different technological environments were examined for two points of time i.e. 1980-81 and 2000-01.

2.5.5.3 Optimal Level of Resource Use

In the past studies, the marginal value productivities from production functions have been used to make recommendations regarding resource shifts to achieve optimum allocation of inputs. Usually, $MVPx_i$ is compared with PX_i. If $MVPx_i$ is less than Px_i, it is suggested that use of resource X_i be curtailed. Such recommendation are based on the principle that profits are maximized under perfect competition when,

$$Py \frac{MPPx_1}{Px_1} = Py \frac{MPPx_2}{Px_2} = Py \frac{MPPx_n}{Px_n} = 1$$

where,

Py = Price of the product
$MPPx_i$ = The marginal physical product of product of ith resource
Px_i = Price of ith resource

The solution is only possible if capital is unlimited, that is, enough capital is available to extend resource use to the optimum, and if all resources are available. It is suggested that such assumptions do not usually hold in practice. If there are resource limitations, only a restricted profits maximization solution can be achieved. The ratio of equation can no longer be equal to unity so that equation now becomes as:

$$Py = \frac{MPPx_1}{Px_1} = Py \frac{MPPx_2}{Px_2} = \quad = Py \frac{MPPx_n}{Px_n}$$

The available limited capital was optimally allocated among different capital inputs appearing in the production function as follows:

The Marginal Rate of Technical Substitution (MRTS) between two inputs was worked out by taking the ratio of marginal products (partial derivatives) of the two inputs. The MRTS between two inputs was then equated to the ratio of the price of these to inputs. Therefore,

$$\text{MRTS of } x_2 \text{ for } x_1 = \frac{MPPx_2}{MPPx_1} = \frac{Y/\ x_2}{Y/x_1} = \frac{Px_2}{Px_1}$$

Thus, several such equations were obtained. The numbers of linearly independent equations were equal to the number of input variables in the equation less one. In fact, there would be as many equations as the number of variables, but one of these equations will be linearly dependent on the other and, therefore, redundant. In addition to the above, one more linear constraint equation was also available, thereby the number of equations becoming equal to the number of unknown input variables. The optimum values of these unknown input variables were obtained by solving these simultaneously. An alternative method could be to maximise a profit function using Langrage Multiplier with the constraint equation. For the production function –

$$Y = ax_1^{b1}\ x_2^{b2}\ x_3^{b3}$$

where,

Y is expressed in rupees and Constraint = $Px_1 . x_1 + Px_2 . x_2 + Px_3 . x_3 = C$

C is the finite amount

The profit function is

$$Z = ax_1^{b1}\ x_2^{b2}\ x_3^{b3} - \lambda\ (Px_1 . x_1 + Px_2 . x_2 + Px_3 . x_3 - C)$$

Setting the partial derivatives of Z with respect to x_1, $x_2\ x_3$ and $\lambda = 0$

We have, λ

$$Z/\partial x_1 = b_1\ ax_1^{b1\text{-}1}\ x_2^{b2}\ x_3^{b3} - \lambda\ Px_1 = 0 \text{ and therefore,}$$

$$b_1\ ax_1^{b1}\ x_2^{b2}\ x_3^{b3} - \lambda\ Px_{1.}\ x_1 = 0 \qquad \text{(i)}$$

Similarly for $\partial Z/\partial x_2$ and Z/x_3

$$b_2\ ax_1^{b1}\ x_2^{b2}\ x_3^{b3} - \lambda\ Px_{2.}\ x_2 = 0 \qquad \text{(ii)}$$

$$b_3\ ax_1^{b1}\ x_2^{b2}\ x_3^{b3} - \lambda\ Px_{3.}\ x_3 = 0 \qquad \text{(iii)}$$

Solving the two equations (i) and (ii)

$$b_1\ b_2\ ax_1^{b1}\ x_2^{b2}\ x_3^{b3} - \lambda\ b_2 Px_{1.}\ x_1 = 0$$

$$b_1\ b_2\ ax_1^{b1}\ x_2^{b2}\ x_3^{b3} - \lambda\ b_1\ Px_{2.}\ x_2 = 0$$

$(b_1 Px_1\ x_2 - b_2 Px_1\ x_1) = 0$, but $\lambda = 0$

Therefore,

$$x_2 = b_2 Px_1 . x_1\ /\ b_1\ Px_2$$

Similarly, $x_3 = b_3\ Px_1 . x_1 / b_1\ Px_3$

Substituting values of x_2 and x_3 in

$$Z/\lambda = Px_1 . x_1 + Px_2 . x_2 + Px_3 . x_3 - C = 0$$

We have,

$Px_1 . x_1 + Px_2 = (b_2\ Px_1 . x_1) / (b_1\ Px_2) + Px_3 = (b_3\ Px_1 . x_1) / (b_1\ Px_3) = C$

Therefore, the optimal quantities of x_1, x_2 and x_3 are,

$$x_1 = \frac{b_1 C}{(b_1 + b_2 + b_3)Px_1}$$

$$x_2 = \frac{b_2 C}{(b_1 + b_2 + b_3)Px_2}$$

$$x_3 = \frac{b_3 C}{(b_1 + b_2 + b_3)Px_3}$$

An alternative method based on the results of Cobb-Douglas production function can be used to estimate the optimum use levels of the resource use under limited situation as below :

$$x_i = \frac{b_i C}{SPx_i} \qquad i = 1,2,3, \ldots, n.$$

where, X_I = Optimum level of ith input

b_i = Production elasticity corresponding to i^{th} resource input

C = Existing total expenditure on all inputs

S = Sum of elasticities

P_{xi} = Price per unit of ith input

2.5.5.4 Share of Technology in Crop Productivity

The share of technology in output was estimated for each crop. The decomposing model (Bisaliah, 1977 and Kiresur et al, 1996) was used to decompose the total productivity differentials between two points of time i.e. 1980-81 and 2000-01. The model requires the Cobb-Douglas production function to be specified on unit area basis. With the help of functional form specified in equation (i), the Cobb-Douglas production function on per hectare basis as specified below, in log linear form for the year 1980-80 was employed.

$$\text{Log } Y_1 = \text{Log } A_1 + a_1 \log H_1 + b_1 \log B_1 + c_1 \log S_1 + d_1 \log M_1 + e_1 \log K_1 + U_1 \qquad \text{......(i)}$$

where,

Y_1 = Output in quintals per hectare for base period.
H_1 = Human labour (mandays) per hectare for base period.
B_1 = Bullock power (pair-days) per hectare for base period.
S_1 = Seed (kg.) per hectare for base period.
M_1 = Manure and fertilizers (Rs.) per hectare for base period.
K_1 = Other working capital (value of irrigation, plant protection, land taxes etc.) in rupees per hectare for base period.
A_1 = Scale parameter
a_1, b_1, c_1, d_1 and e_1 are the output elasticites of respective inputs.

Similarly, per hectare production function for the year 2000-01 was,

$$\text{Log } Y_2 = \text{Log } A_2 + a_2 \log H_2 + b_2 \log B_2 + c_2 \log S_2 + d_2 \log M_2 + e_2 \log K_2 + U_2 \quad .....(ii)$$

The definitions of variables and parameters in equation (ii) are as in equation (i).

The decomposition analysis proceeds by taking the difference of the linearised production functions, using geometric mean values of each variables. Taking differences between equation (ii) and equation (i) and adding same terms and substracting the same terms we get,

$$\text{Log } Y_2 - \text{Log } Y_1 = (\text{Log } A_2 - \text{Log } A_1) + (a_2 \log H_2 - a_1 \log H_1) + (b_2 \log B_2 - b_1 \log B_1) + (c_2 \log S_2 - c_1 \log S_1) + (d_2 \log M_2 - d_1 \log M_1) + (e_2 \log K_2 - e_1 \log K_1) + (U_2 - U_1) \quad ....(iii)$$

where,

Y_2 and Y_1 = Geometric mean yields of the new and old period
A_2 and A_1 = Estimated intercepts of equation (i) and (ii)
a_i and b_i = Estimated elasticities of inputs during old and new period.
H_j, B_j, S_j, M_j, and K_j = Geometric mean levels of inputs

Rearranging terms in equation (iii) we get,

$$\text{Log } Y_2 - \text{Log } Y_1 = [\text{Log } A_2 - \text{Log } A_1] + [(a_2 - a_1) \log H_1 + (b_2 - b_1) \log B_1 + (c_2 - c_1) \log S_1 + (d_2 - d_1) \log M_1 + (e_2 - e_1) \log K_1] + [a_2 (\log H_2 - \log H_1) + b_2 (\log B_2 - \log B_1) + c_2 (\log S_2 - \log S_1) + d_2 (\log M_2 - \log M_1) + e_2 (\log K_2 - \log K_1) + [U_2 - U_1] \quad ...... (iv)$$

Equation (iv) can also be written as,

$$\begin{aligned}\text{Log}\,[Y_2/Y_1] = \text{Log}\,[A_2/A_1] &+ [(a_2 - a_1)\,\text{Log}\,H_1 \\ &+ (b_2 - b_1)\log B_1 + (c_2 - c_1)\log S_1 \\ &+ (d_2 - d_1)\log M_1 + (e_2 - e_1)\log K_1\,] \\ &+ [a_2 \log (H_2/H_1) + b_2 \log (B_2/B_1) \\ &+ c_2 \log (S_2/S_1) + d_2 \log (M_2/M_1) \\ &+ e_2 \log (K_2/K_1)] + [U_2 - U_1] \qquad \text{.......(v)}\end{aligned}$$

The decomposition equation (v) involves decomposing the natural logarithm of the ratio of 'new' output to 'old' and approximates the percentage change in output with new technology holding all other factors constant. The left hand side of the equation gives the total difference in productivity expressed as percentage over old period technology. The first bold bracketed expression on the right hand side (the neutral logarithm of ratio of intercept term), is a measure of percentage change in output due to shift in scale parameter (A) of the production function, i.e. neutral technological change; the second bold bracketed expression, the sum of the arithmetic changes in output elasticities each weighted by the logarithm of the volume of the quantity of a particular input used, is a measure of percentage change in output due to shifts in slope parameters (output elasticities) of production function, i.e. non-neutral technological change; the third bold bracketed expression is the sum of logarithm of the ratio for each input of 'new' and 'old' input, each weighted by the output elasticity of that input. This expression is a measure of percentage change in output due to change in the per hectare quantities of human labour, bullock power, seed, manure and fertilizers, and other working capital used, given the output elasticities of these inputs under new production technology. The first two expressions can be summarized up and interpreted as change in yield to technological change i.e. because of adoption of technology. The third term may be interpreted as the change in crop yield due to inputs used.

The structural difference between two functions is tested by using 'Chow's test' (Koutsoyiannis, 1977). The Chow test is constructed as 'F' ratio, as follows:

$$F = \{[S_3 - (S_1+S_2)] / n - (n_1-n_2)]\} / \{(S_2) / (n_1+n_2)\}$$

where,

'S_1' and 'S_2' are residual sum of squares for the two response functions, i.e. old and new technology functions.

'S_3' is the residual sum of squares of the pooled regression model (old and new technology combined)

'n_1' and 'n_2' are degrees of freedom for old and new technology functions.

'n' is degrees of freedom for the pooled data.

This test identifies whether there are structural difference between the two equations, but does not ascribe this difference if it exists, to a) intercept difference, (b) difference in slope or (c) both.

2.5.5.5 Sustainability Analysis

The sustainability of new technology could be reflected by its proportional response to the existing crop situations, which in turn could be quantified by means of a site-specific index (which may be called as 'Site Index'). For a given crop growing situations, the site index was computed for each selected farmer of old period i.e. 1980-81, by taking derivatives of the yield per unit area from overall mean of all such sample fields. The regression of the yield per unit area in fields of new period i.e. 2000-01, on the corresponding site indices give a measure of response of new technology with regard to the level of site indices, which in turn represents the yield of old period.

The economic sustainability of crop production technologies was assessed using Sustainability Index (Kiresur et al, 2001) in terms of yields. This method of estimating the desirable index of improved agricultural technologies for their economic sustainability of yield was used with slight modifications as below. The similar formula was employed to compute sustainability index for net returns also.

Let there be 'n' farms selected for 2000-01 and let Y_{2i} (i = 1 to n) be the yields of ith farm and Y_{2m} be the mean yield of all such farms in a given crop growing situation.

Let Y_{1i} (i = 1 to n) be the corresponding yields during the year 1980-81 and Y_{1m} be the mean yield of the farms. Then for each field, a site index (E_i) is computed as,

$$Ei = (Y_{1i} - Y_{1m})$$

$$\text{Then, } Y_{2i} = a + b E_i$$

where,

Y_{2i} = Yield level in 2000-01

a = Constant (intercept)

b = Regression coefficient

Let 'b' be the estimated regression coefficient of Y_{2i} on Ei. Then adopting the linear function method (Hernandez et al., 1993) the desirable yield of technology adoption for the ith farm was worked out as:

$$Di = Y_{2i} + b\,S,$$

where,

Di = Desirable yield level due to technology during 2000-01 for i[th] field

S = (E max + E min) / 2

S = Symmetry of Site Indices

E max = Maximum value of Ei

E min = Minimum value of Ei

The E max and E min are the maximum and minimum values of the site indices for given crop. In fact the 'S' reflects the symmetry of the site indices. The standard yield level (Ds) attained by the new technology (2000-01) for a particular crop was defined as :

$$Ds = Y_{2m} + S.$$

Sustainability Index (SI) = Di >Ds / n

where, n = sample size

By computing Di and Ds values, the expected yield level obtained due to new technology were tested against standard yield level (Ds) and sustainability of technology was evaluated as the percentage of those Di's value which were statistically on par or greater than Ds values. The significance of the difference between Di and Ds was tested with the help of 't' test using the following equation.

$$t_{n-2} = | Di - Ds | / SE\,(Di)$$

The standard error of Di was computed as :

$$S.E._{(Di)} = \sqrt{(1/n + S^2 / \Sigma Ei^2)}\ MSE$$

where,

n = Number of farms
S = Symmetry of site indices
Ó Ei² = Sum of squares of site indices for given group
MSE = Mean square error of yield obtained due to adoption of technology in the analysis of variance

2.6 Definitions and Cost Concepts Used in the Study

2.6.1 Evaluation of Items of Cost

(i) Farm land : - Self cultivated land evaluated at prevalent rates in the village, taking in to account the differences in type of soil, distance of village, source of irrigation availability etc. The land improvement done later was evaluated on the basis of actual cost.

(ii) Dwelling house, cattle shed, storage sheds, wells, etc.:- Evaluated at prevailing prices in the village.

(iii) Implements and machinery :- Evaluated at market prices (Annual replacement of minor implements treated as current expenses).

(iv) Manure and fertilizers : - Farm produced manure evaluated at prevailing market prices in the village. In case of purchased manure and fertilizers, the actual cost of purchase was accounted for.

(v) Farm output : - Both main-produce and by-produce evaluated at farm harvest prices prevailing in the village.

(vi) Casual hired labour : - Actual amount paid in cash and kind, the later evaluated at the harvest prices, prevailing in the village. Prerequisite if any, were included in kind payments and evaluated at market prices.

(vii) Attached labour : - Actual amount paid in cash and kind, the later evaluated at the harvest prices, the payments in kind includes food, clothes and either supplies including prerequisites.

(viii) Family labour : - It is imputed on the basis of wage rate for casual labour prevailing during the period of operation.

(ix) Bullock power : - The hired bullock power was evaluated at actual amount paid in cash or kind, the later evaluated at the harvest prices or retail prices prevailing in the village while, the owned bullock power was evaluated on the basis of maintenance cost of bullocks.

(x) Depreciation of farm buildings, implements and machinery:- The depreciation of farm buildings, irrigation structures, implements, machinery, hand-tools etc. was computed by using the straight line method. (i.e. Present value less junk value divided by expected remaining life in years)

(xi) Interest on fixed capital :- The interest on fixed investment on different farm assets was calculated at the rate of 10 per cent per annum.

(xii) Irrigation charges :- If own irrigation facility were available, the irrigation cost (IC) was worked out as:

$$IC = D + R + S / GIA \times P + [(H \times M)]$$

where,

D = Depreciation on irrigation structures in rupees.
R = Repairs on irrigation structures in rupees.
S = Interest on fixed investment on irrigation structure in rupees
(Excluding well, tube wells).
P = Plot area in hectare
H = Machine hours utilized to irrigate the plot area.
GIA = Gross irrigated area in hectare
M = Per hour maintenance cost of the machine in rupees.

Per hour maintenance cost of machine is calculated as the total expenditure on oil, lubricants, electricity charges, minor repairs, divided by the total working hours of the machine. The charges of hired machine were evaluated at the prevailing hiring charges in the village.

(xiii) Rental value of land :- It was computed as one sixth of the value of gross produce less land revenue and other taxes.

(xiv) Working capital :- It includes wages of hired labour, wages of hired and owned bullock power, hired and owned machinery charges, seed value (both farm produced and purchased), value of manure, fertilizers, insecticides and pesticides, irrigation charges and crop insurance premium.

(xv) Interest on working capital :- It was charged at the rate of 14 per cent per annum for the life period of the crop.

(xvi) Incidental charges:- The cost of acquisition of inputs was treated as incidental charges.

(xvii) Per quintal cost of production :- It is calculated by taking per hectare total cost of cultivation (i.e. Cost- C) divided by quantity of main produce in quintal.

2.6.2 Cost Concepts Used

Basically, the various costs of cultivation are divided into three categories viz., Cost- A, Cost- B and Cost- C. These cost concepts were adopted, while studying the cost of cultivation of crops.

(i) Cost- A:- It includes cost items of working capital plus interest on working capital, depreciation on implements, machinery and farm buildings, plus land revenue and other taxes.

(ii) Cost- B :- It comprises of Cost- A plus imputed rental value of owned land and imputed interest on owned fixed capital (excluding land).

(iii) Cost- C :- It includes Cost-B plus imputed value of family labour.

3

Agricultural Profile of Maharashtra

The purpose of this chapter is to discuss in brief the geographical features and agricultural profile of Maharashtra State. The information relative to the different agro-climatic zones of Maharashtra State is, in effect, an integral part of the present investigation, as it permits easy understanding of the study results in their right perspective.

3.1 Geographical Features of Maharashtra

Agriculture is the main source of livelihood for about 63 per cent population of Maharashtra. The share of agriculture in the State income was about 13.6 per cent (2001-02). Apart from providing food for the population, agriculture supplies raw material and finished goods for exports, thereby helping the country to earn the much-needed foreign exchange. The factors such as soil types, topography, weather conditions, etc; are important ones directly influencing the crop growth, besides the non-conventional inputs such as seeds, fertilizers, irrigation and entrepreneurship etc. The study of these factors together with land use pattern and crop pattern would help in understanding the salient features of the State.

3.1.1 Location

The State of Maharashtra has been formed on the first day of May, 1960. The State forms a major part of the peninsular India. Maharashtra State is the third largest State in India in terms of area and population. The total area of the State is 3.08 lakh sq. km. It accounts for nearly 10 per cent of the total area of the Indian Union.

The State extends over 800 km from East to West and about 700 km from North to South and lies between 16.4° to 22.1° North latitudes and

72.6° to 80.9° East longitudes. The State is flanked by the Arabian Sea at the west by about 720 km coastal area. On the West, it is bounded by Arabian Sea, Gujarat and Goa, and part of the Union Territory viz., Dadra and Nagar Haveli to its North-West, Madhya Pradesh in the East and North, Andhra Pradesh in South-East and Karnataka and Goa in the South.

3.1.2 Topography

The main features of Maharashtra's topography are the famous North-South range of the Western Ghats called the *'Sahyadri'* running almost parallel to the sea coast on the West and East-West ranges of hills called the *'Satpuda'* on the Northern border of the State. There are also smaller hilly ranges and high lands to the North of Nasik district called the '*Satmala*' ranges with their continuation in the '*Balaghat*' range which goes right upto the Eastern boundary of Vidarbha. The Eastern Ghats divide the State into two distinct natural parts called the Konkan and the Deccan plateau. The Konkan is a long and narrow hilly undulating region of 40 km to 60 km width lying between the Arabian sea and the Western Ghats stretch the long and broad plains or table lands well recognised as the Deccan plateau at an elevation ranging from 450 to 610 metre above the sea level. The plateau is intercepted by *Satpuda, Balaghat* and *Mahadeo* hills, which give rise to three fertile river basins called the Tapi, Godawari and Krishna basins. The names are derived from the major rivers flowing through them.

3.1.3 Soils

Of the five factors of soil formation and development, the climate and topography have predominantly influenced the soil formation in Maharashtra. The main geological formation in the entire State is the Deccan Trap or basalt. Its laterisation is, however, noticed in Ratnagiri and Sindhudurg districts of Konkan region. The far eastern region of Chandrapur, Gadchiroli and Bhandara districts have various rock formations like, granite, gneiss, schist and trap. The natural features have given rise to the following kinds of soils.

A. Black Soils

These are most important soils covering the maximum area in the State. They are derived from Deccan Trap and are found in all districts, excepting Ratnagiri and Sindhudurg districts and parts of Chandrapur district. These Soils have fairly good potash content, fair to less than fair phosphates and a low nitrogen and organic matter content. These are clayey soils and clay content varies from 30 to 60 per cent and above. The pore space varies from 40 to 60 per cent. They are mildly alkaline in reaction with pH varying from 7.2 to 8.5. These soils are classified as under.

(i) The shallow black soils : These soils are lighter or dull-grayish black in colour and their depth is about 30 cm. they are usually quite low in fertility.

(ii) The medium black soils : These soils are more clayey than the above and their depth is around a metre. Their black colour becomes somewhat lighter down the profile. The total exchangeable bases are very high, varying from 50 to 75 mc. of which exchangeable calcium alone constitute nearly 60 to 80 per cent.

(iii) The deep black soils : These soils are mainly developed on alluvial material along the courses of big rivers like Tapi, Godawari, Bhima, Krishna, etc., and are observed on low lying or flat topography. These are highly fertile soils.

B. Red Soils

These soils are observed in Thane, Raigad and in the western portion of Pune district. In the East, they are noticed in Chandrapur, Gadchiroli and Bhandara districts. The later soils are formed from mixed parent material such as granite, gneiss, schist, trap, etc. on hill slopes, the soils are coarse sandy in nature; while in valleys, they are like clay loams. The soils of the hill slopes may be only 9 cm to 45 cm in depth, while those of the flat lands and valleys are deeper than 60 cm. Their pH is slightly on the acidic side.

C. Lateritic Soils

On account of the forest vegetation and warm humid climate, the Deccan trap has given rise to laterites and lateritic soils in the Southern part of Raigad, most of Sindhudurg, Ratnagiri and Western part of Satara, Sangli, Nasik and Kolhapur districts. As a result of alkaline reaction of the percolating waters, the silica from the original rock is differentially leaving a residue of sesqui-oxides. Intense leaching accompanying the high rainfall has removed from soil all the bases, giving it a distinctly acidic character with a pH 5.00 to 6.00. There is completely absence of calcium carbonate and to a great extent of available phosphorus and potash. The soils are usually loamy in texture, fairly deep but with low retentively of moisture. They are well drained due to which inspite of the heavy rainfall, there is scarcity of water in the summer months. The laterites have three varieties of soils, viz. rice soils, *warkas* soils and garden soils. The rice soils are generally in low-lying areas, which receive the washings from the upland forest and are usually rich in organic matter. The *warkas* soils which are on hill slopes are shallow and eroded soils, coarse in texture and yellowish red in colour. These soils respond well to N and P fertilizers. The garden soils are light, easily workable, yellowish to brownish in colour and mostly support areca nut and coconut plantation.

D. Alluvial Soils

These soils are of two types viz., (a) alluvial along the main rivers like Tapi, Godawari, Bhima, Krishna, etc., and (b) coastal alluvial as noticed on the coastal side of Thane, Ratnagiri and Sindhudurg districts. The former soils are more clayey; while later are more silty and sandy in nature. These are very fertile soils.

E. Forest and Hill Soils

These soils are also derived from trap and occupy the mountainous terrain in the west of the districts of Nasik, Ahmednagar, Pune, Sangli and Kolhapur; and in the east of Gadchiroli district. As the rainfall in this region is comparatively high, the soils are low in base status and neutral to slightly acid in reaction. These are also known as *kumris, bardi, retadi,* etc., and thus indicate their coarseness.

F. Saline-alkali Soils

These soils are of two types viz., (a) coastal saline soils and (b) saline-alkali and alkali soils of inland region. The coastal saline soils occur all along the west coast in the districts of Ratnagiri, Sindhudurg, Raigad and Thane. The soils, which get ingress of seawater, are impregnated with salts to the extent of 0.2 per cent of which nearly 50 per cent is NaCl from sea. Inspite of heavy rainfall, they are not capable of growing any sweet land varieties of paddy. These are locally known as *khajan* lands.

The saline-alkali and alkali soils of the inland areas are having degree of sodium saturation and nearness of water table. The vast acreage of such soils are met with in the Deccan canal tract. These soils are incapable of normal growth of crops unless suitable ameliorative measures are adopted to reclaim them.

3.1.4 Climate and Rainfall

The State can broadly be divided into four meteorological sub-divisions viz., Konkan, Western Maharashtra, Marathwada and Vidarbha. The State experiences three distinct seasons as Southwest Monsoon Season (*Kharif*) during June to September, Post-Monsoon or Winter

Season (*rabi*) from October to February and Summer Season in March, April and May. The Southwest Monsoon is the main rainy season during which different parts of the State receive the major portion of the precipitation. The winter rains have relatively little significance and account for at the most 5 to 6 per cent of the rainfall received by the eastern districts of the Western Maharashtra and the western districts of the Marathwada region.

The annual rainfall varies from 2250 to 2540 mm in coastal belt. It suddenly jumps to over 6350 mm at certain locations like Mahabaleshwar and Amboli on the crest of Sahyadri hills. On the eastern flanks of Sahyadri, also known as Maval and transition tract, the rainfall varies from 1270 to 2540 mm. A zone of low rainfall then follows this region. The annual rainfall of less than 750 mm, is unstable, unevenly distributed and of erratic nature. The drought prone area covers nearly one-third of the total area of the State. The low and erratic rainfall and dry spells of varying duration are usually experienced in July and August. Generally, two dry spells of more than 4 weeks duration result in failure of *kharif* crops grown between June-October. Besides these usual features, weather aberrations such as delay onset of monsoon, timely onset followed by dry spells and early withdrawal of monsoon by September are of common occurrence. All these aberrations affect crop production considerably.

As the monsoon currents approach the high land plateau of the Vidarbha region, they start precipitating the residual moisture so that the eastern districts of the Marathwada region and the entire Vidarbha region receive relatively assured rains, which in the eastern border districts of Chandrapur, Gadchiroli and Bhandara exceed 1524 mm per annum.

During the study year i.e 2000-01, the light to medium pre-monsoon rains were received in all parts of the State during the month of May. Rains received were found beneficial for preparatory tillage operations of *kharif* crops. Monsoon commenced in south Konkan on 4th June and reached in rest parts of the State upto 7th June. The monsoon rains received were above normal in 13 districts, 75 to 100 per cent of the normal rainfall in 12 districts and 50 to 75 per cent of normal in 9 districts. In the State, the *kharif* sowing was completed to the extent of 102 per cent. In general, the condition of *kharif* crops reported to be satisfactory, but yields of all *kharif* crops went down to some extent due to short dry spell. The State experienced widespread and satisfactory rains on the onset of *rabi* season but crop-stand of *rabi* crops during later stage was affected due to long dry spell. The sowing of *rabi* crops was to the extent of 87 per cent. During the year 2000-01, the State received average annual rainfall of about 1188 mm, which was less by 3.85 per cent than that of normal rainfall (1233.28 mm).

On the basis of temperature, the state can be divided into four main regions. The first region comprises the coastal districts, which have low annual variation in temperature. The winter in this region remains very mild and the humidity relatively high. The second region comprising the districts of Western Maharashtra, is characterized by relatively hot summer, humid and temperate humid rainy season and moderately cold winters. The third region comprising the districts of Marathwada and the fourth region comprising the districts of Vidarbha differ from each other only in degree in so far as the temperatures are concerned. The temperatures in the Vidarbha are relatively higher than in the Marathwada region. In both the regions, the winter temperature is rather low. The maximum and minimum temperatures in the State vary between 26° to 44°C and 8° to 27°C, respectively. The coverage of all the four regions is depicted in Appendix-I.

3.1.5 Agro-climatic Zones of Maharashtra

Maharashtra because of its size, topography and varying soil types, rainfall patterns and temperatures is divided into different agro-climatic zones. As a matter of fact, the vast area of 3.08 lakh sq km with such varying soils, water and climatic features are bound to have different agro-climatic conditions. The state is traditionally divided into 16 zones by National Commission on Agriculture in 1965 based on mainly rainfall pattern with adaphic topographic and detailed crop pattern. However, these zones are not agro-climatically homogenous. Therefore taking into consideration the factors such as, nature of soil, predominant cropping pattern, major natural vegetation and their types, hypsometrical (latitudinal) situation, total annual rainfall, weekly distribution pattern of rainfall, mean daily maximum and minimum temperatures, mean daily per cent relative humidity at morning and afternoon and mean monthly evaporation index, the State is divided into nine agro-climatic zones. The detailed information of nine agro-climatic zones in Maharashtra and their coverage is depicted in Appendix- II and III, respectively.

Zone I : Very high rainfall zone with lateritic soils

This zone consists of whole of Ratnagiri, Sindhudurg districts and western part of six tahsils from Kolhapur district. The general topography is hilly to undulating and soil depth varies from 30 cm on hill slopes to 1.2

metres in valleys. The soils in low laying valleys are red and reddish brown, while soils of higher slopes are yellowish red. The texture is sandy loam to clay loam. Soils are coastal alluvial and coastal saline. The climate is warm and humid with more than 2500 mm of rainfall. The minimum and maximum temperatures range between 22° C and 24°and 30° C and 31° C, respectively.

Zone II : Very high rainfall zone with non-lateritic soils

It covers whole Thane, Greater Mumbai and Raigad districts. The western parts of three tahsils from Nasik district come under this zone. The natural vegetation is more. The general topography is hilly to undulating, but mostly less than 300 metres above mean sea level. Soils derived from Deccan trap are non-lateritic and are formed under condition of leaching. The soil depth varies from 30 cm to one metre. The soils are coastal alluvial and coastal saline. The rainfall is more than 2250 mm during the year.

Zone III : Ghat zone

It includes the hilly, high lying terrain round about the ghats of Amboli and Fonda of Konkan and parts of Kolhapur, Satara, Pune, Nasik districts and Akola tahsil of Ahmednagar district. The soils are shallow light, brown to dark brown on hilly slopes, red to reddish brown lateritic soils on high level western parts. The highest rainfall is obtained in this zone.

Zone IV : Transition zone - A

It comprises western hilly flank of Kolhapur, western part of Shirala tahsil of Sangli, central strip of Patan and Jaoli tahsils of Satara, western part of Bhor, Velhe and Paud tahsils as well as central part of Vadgaon tahsil and western strip of Khed, Ambegaon and Junnar tahsils of Pune district, eastern part of Igatpuri, central part of Trimbak and Peint, western Dindori and western half Surgana tahsils of Nasik district and western side of Akole tahsil in Ahmednagar district. The soils are red to brown with varying depths and textures. The annual precipitation ranges between 1750 to 2500 mm.

Zone V : Transition zone - B

It is wider strip running parallel to Transition zone A on its eastern side. The zone extends towards east upto the line where scarcity zone starts. It comprises eastern strip of Kolhapur, eastern part of Shirala of

Sangli, central strip of Satara district, central part of Bhor and western Poud, eastern part of Junnar, Haweli, Abmegaon tahsils from Pune district, central portion of Akole tahsil of Ahmednagar district, western Sinnar, eastern Igatpuri, central portion of Nasik, eastern part of Satana tahsil in Nasik district, western part covering of Akkalkuwa and Akrani, and western part of Taloda, Nandurbar and Sakri tahsils in Dhule district. Soils are predominantly grayish black with varying texture and depth. The rainfall ranges between 700 to 1750 mm.

Zone VI : Scarcity zone

This zone comprises of Shirol from Kolhapur, entire Sangli, Satara, Pune, Ahmednagar, Nasik and Dhule districts (excluding the portion already mentioned above), western strip of Beed, whole of Solapur district excluding eastern Barshi, south Solapur and whole Akkalkot, Paranda and Bhoom tahsils of Osmanabad district, Part of Aurangabad district and one tahsil from Jalana district. The climate is bounded by isohyets of 700 mm on both sides. There are places receiving even 500 m of rain annually. Soils are calcareous grayish black in colour and of varying depth and texture.

Zone VII : Assured rainfall zone

It covers remaining parts of Solapur, Osmanabad, Latur, Beed, Aurangabad, Jalgaon districts; whole of Buladhana, Akola, Amarawati districts; half of Parbhani district and southern Nanded district and part of Yeotatmal district. The topography is almost flat to rolling. The soils are more or less similar to VI zone. However, clayey deep soils have a problem of temporary water logging. The average rainfall received is 700 mm on the western side and upto 900 mm on eastern side.

Zone VIII : Moderate to Moderately high rainfall zone

This zone covers remaining parts of Nanded, Parbhani, Akola, Yeotmal and Amrawati districts, whole of Wardha, Nagpur (excluding the western strip) and western part of Chandrapur district. The rainfall with 900 mm on western side increases upto 1250 mm on the eastern side. The soils are calcareous and grayish black in colour and varying depth and more fine in textures. The problem of temporary water logging also exists in this region. The rainfall ranges between 900 mm to 1250 mm.

Zone IX : High rainfall zone with formed from mixed rocks

This zone covers whole Bhandara district and remaining parts of Nagpur and Chandrapur, and Gadchiroli district. It forms extreme eastern part of State. The rainfall ranges between 1250 mm to 1700 mm.

3.2 Agricultural Profile of Maharashtra

The present study is based on the information collected at two points of time with time lag of 20 years. During this period the agriculture in the different agro-climatic zones of Maharashtra was subjected to various (agricultural) development programmes formulated and administered under the multilevel planning approach, which enjoyed wide acceptance in planning for economic development. The strategy adopted for this purpose was expected to bring about the desirable influence at the farm level, as it included a resource development approach based on the generalization of the scientific management and development of productive resource for sustained increase in the levels of living of the community as a whole. However, because of wide range of diversity in resources, climatic conditions and institutional set-up, the agriculture economy is inflicted with economic and technological dualism indicating the differences into the agro-climatic zones.

In fact, the government made various attempts by launching different programmes to increase agricultural output for the several years in different agro-climatic zones. The agriculture of these zones, however, exemplifies the complexity of the agricultural development process in an underdeveloped area. The heterogeneity of the physical environment among the zones indicated below is largely responsible for this complexity.

3.2.1 Population

According to 1981 census, the population of the State was 62.8 million constituting about 10 per cent of the population of India, as against 96.75 million according to population census of 2001. The population of the State has increased by 54.06 per cent during the last two decades. The distribution of the population in different agro-climatic zones for the year 1981 and 2001 are presented in Table 3.1, so also graphically indicated in Chart-1.

Table 3.1: Agro-climatic Zonewise Population in Maharashtra

Zones	*1980-81*		*2000-01*		*Percentage*
	Population (Million)	*% to total*	*Population (Million)*	*% to total*	*change over*
I	2.00	3.18	2.85	2.95	42.56
II	4.80	7.64	22.25*	23.00	363.54
III	1.10	1.75	1.17	1.21	6.61
IV	3.90	6.21	5.67	5.86	45.33
V	4.40	7.01	5.56	5.75	26.46
VI	20.40	32.48	23.02	23.79	12.85
VII	16.70	26.59	18.88	19.52	13.08
VIII	5.70	9.08	12.27	12.68	115.26
IX	3.80	6.05	5.07	5.24	33.48
Maharashtra	62.80	100.00	96.75	100.00	54.06

Note- * indicates the population including Mumbai city and Mumbai suburban district

Source :- 1. Population census for the year 1981.
2. Human Development Report, Maharashtra- 2002. Govt. of Maharashtra.

Chart 1 : Changes in Population in Different Agro-Climatic Zones of Maharashtra

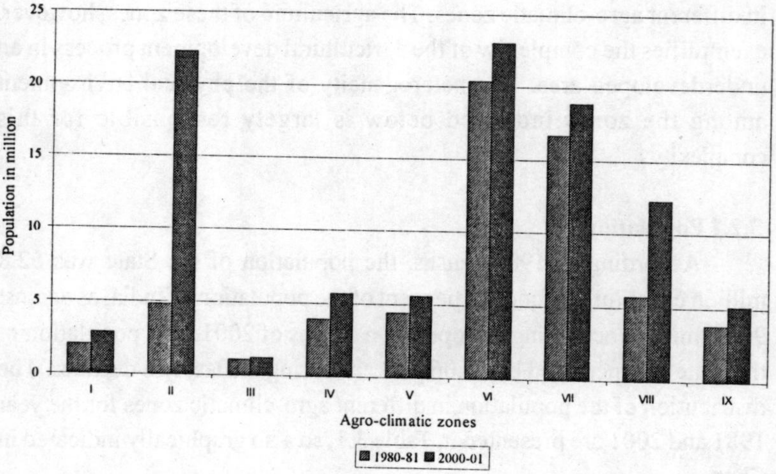

The percentage of rural population has increased from 40.7 per cent in 1981 to 57.6 per cent in 2000-01 of the total population of the State. The literacy percentage has substantially increased from 47 per cent in 1981 to 77.3 per cent in 2001. The persons engaged in various pursuits of agriculture from the bulk of working population has somewhat declined from 64.88 to 62.28 per cent, whereas, the density of population has increased from 204 persons per sq. km to 314 persons per sq. km during the period of 20 years. The decadal population growth rate and the sex ratio of the State declined to 22.6 per cent from 24.54 per cent and 937 to 922 per thousand male, respectively, during the period under study. (Economic Survey of Maharashtra- 2001-02)

During the year 1981, the share of population of an individual zone in the total population of the State, was higher in zone VI (32.48 %) followed by zone VII (26.59 %), while, it was in the range of 5 to 10 per cent in zones II, IV, V, VIII and IX. The population share was below 5 per cent in the zones I and III. The same trend was also observed in the year 2001. However, the percentage increase in the population over the period was observed higher in zone VIII (115.26 %). Next to this, the zones IV, I and IX were ranked with increment in the population over a period in the order of 45.33, 42.56 and 33.48 per cent, respectively.

3.2.2 Land Use Pattern

The information relating to land use pattern of the individual agro-climatic zones and Maharashtra state as a whole for the year 1980-81 and 2000-01 alongwith the percentage change in individual land use categories in the total geographical area of the respective agro-climatic zones is presented in Table 3.2, alongwith the graphical presentation in Chart-2.

In the year 1980-81, with the net-cropped area of 180.08 lakh hectares and gross cropped area of 196.42 lakh hectares, Maharashtra ranked second in net area sown and third in gross cropped area accounting for 13 and 12 per cent of the respective areas in India.

During the period under reference i.e. 1980-81 to 1998-99, the total geographical area in the State has remained almost the same i.e. 307.58 lakh hectares. The proportion of net-cropped area, which was already as high as 58.56 per cent (180.08 lakh hectares) of geographical area, has been declined by 1.82 per cent over a period of 20 years.

Table 3.2 Change in Land Use Pattern of Different Agro-climatic Zones in Maharashtra State

Agro-climatic zones	*Geographical area*	*Area under forest*	*Land under non-agril. use*	*Barren and uncultivable land*	*Permanent pasture and grazing land*	*Land under misc. trees not included in net area sown*	*Cultivable waste*	*Current fallow*	*Other fallow*	*Net cropped area*
Zone I : 1980-81	12.79	1.38	0.33	3.42	0.17	0.53	2.93	0.50	1.17	3.36
2000-01	12.79	0.51	0.37	3.37	0.19	0.39	2.04	0.43	2.04	3.86
% change over		-63.04	12.12	-1.46	11.76	-26.42	-30.38	-14.00	74.36	14.88
Zone II : 1980-81	16.59	6.41	1.22	2.17	1.27	0.30	0.86	0.41	0.57	4.46
2000-01	16.59	6.09	1.52	1.92	1.05	0.27	0.72	0.24	0.65	5.33
% change over		-4.99	24.59	-11.52	-17.32	-10.00	-16.28	-41.46	14.04	19.51
Zone III :1980-81	4.50	1.19	0.36	0.52	0.18	0.01	0.08	0.09	0.06	2.01
2000-01	4.50	0.77	0.40	0.39	0.21	0.03	0.11	0.08	0.11	2.34
% change over		-35.29	11.11	-25.00	16.67	200.00	37.50	-11.11	83.33	16.42

(Contd.)

(Contd.)

Agro-climatic zones	Geographical area	Area under forest	Land under non- agril. use	Barren and uncultivable land	Permanent pasture and grazing land	Land under misc. trees not included in net area sown	Cultivable waste	Current fallow	Other fallow	Net cropped area
Zone IV : 1980-81	16.44	2.83	0.70	1.41	0.82	0.12	1.90	0.56	0.93	7.19
2000-01	16.44	3.21	0.61	1.22	0.57	0.08	0.38	0.31	0.67	7.98
% change over		13.43	-12.86	-13.48	-30.49	-33.33	-80.00	-44.64	-27.96	10.99
Zone V : 1980-81	16.93	2.63	0.37	1.08	0.85	0.02	0.15	0.24	0.20	8.79
2000-01	16.93	1.82	0.46	0.91	0.62	0.08	0.32	0.35	0.42	7.63
% change over		-30.80	24.32	-15.74	-27.06	300.00	113.33	45.83	110.00	-13.20
Zone VI : 1980-81	90.15	4.56	2.46	5.41	3.14	0.36	2.02	2.09	3.73	62.94
2000-01	90.15	4.29	2.37	4.20	3.09	0.47	1.80	4.09	3.95	60.16
% change over		-5.92	-3.66	-22.37	-1.59	30.56	-10.89	95.69	5.90	-4.42
Zone VII : 1980-81	73.60	5.98	2.01	1.14	2.72	0.51	1.18	3.48	2.04	54.54
2000-01	73.60	7.17	2.41	0.91	2.61	0.36	1.71	3.62	2.16	50.62
% change over		19.90	19.90	-20.18	-4.04	-29.41	44.92	4.02	5.88	-7.19

(Contd.)

(Contd.)

Agro-climatic zones	*Geographical area*	*Area under forest*	*Land under non- agril. use*	*Barren and uncultivable land*	*Permanent pasture and grazing land*	*Land under misc. trees not included in net area sown*	*Cultivable waste*	*Current fallow*	*Other fallow*	*Net cropped area*
Zone VIII : 1980-81	33.75	7.91	0.85	0.78	1.32	0.21	0.32	0.65	0.50	24.15
2000-01	33.75	6.08	1.66	0.71	1.79	0.34	0.82	1.31	0.67	25.28
% change over		-23.14	95.29	-8.97	35.61	61.90	156.25	101.54	34.00	4.68
Zone IX : 1980-81	42.83	20.40	2.18	1.44	5.29	0.06	0.48	0.56	0.78	12.64
2000-01	42.83	21.57	2.59	0.78	3.27	0.20	0.98	0.89	0.71	13.61
% change over		5.74	18.81	-45.83	-38.19	233.33	104.17	58.93	-8.97	7.67
Maharashtra:1980-81	307.58	53.29	10.48	17.37	15.76	2.12	9.92	8.58	9.98	180.08
2000-01	307.58	51.51	12.39	14.41	13.40	2.22	8.88	11.32	11.38	176.81
% change over		-3.34	18.23	-17.04	-14.97	4.72	-10.48	31.93	14.03	-1.82

Chart 2: Changes in Land Use Pattern of Maharashtra During 1980-81 to 2000-01

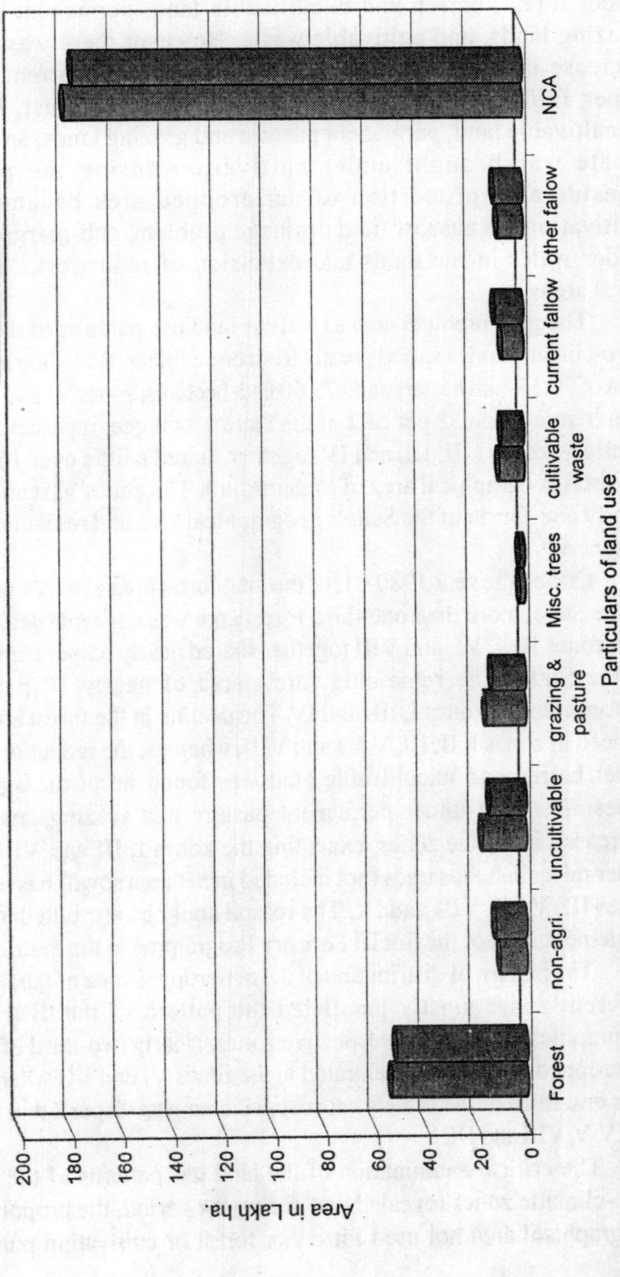

During this period, there was considerable reduction in area under forest, barren and uncultivable land, permanent pasture and grazing lands, and cultivable waste. However there was remarkable increase in the land under non-agricultural use, current fallow and other fallow land. Although, the part of the forest, barren and uncultivable land, permanent pasture and grazing lands, and cultivable waste was brought under cultivation during the period, the considerable proportion of net cropped area became unfit for cultivation because of field drainage problem, sub-mergence of land under water in the dams and extension of road works, building in rural areas.

The geographical area as well as land use patterns of the individual agro-climatic zones varied greatly from one another. With their geographical area of 90.15 lakh hectares and 73.60 lakh hectares, zones VI and VII together shared more than 53 per cent of the State's total geographical area. On the contrary, zones I, II, III, and IV together shared a little over 16 per cent of the total geographical area of Maharashtra. The zones VIII and IX shared nearly one-fourth of the State's geographical area and remaining area falls in zone V.

During the year 1980-81, of the total forest area of 53.29 lakh hectares in the State, more than one-third forest area was concentrated in zone IX. The zones II, V, VI, and VIII together shared nearly 55 per cent forest area in the State. The remaining forest area of nearly 10 per cent was concentrated in zones I, III, and IV. The decline in the forest area has been noticed in zones I, II, III, V, VI and VIII, whereas, the reduction in the area under barren and uncultivable land was found in all the agro-climatic zones. The area under permanent pasture and grazing land was also decreased in all the zones, excepting the zones I, III and VIII. The land under miscellaneous trees (not included in net area sown) has increased in zones III, V, VI, VIII and IX. The reason could be attributed to the better implementation of the Social Forestry Programme in the State.

The pattern of distribution of the net-cropped area of the State among different zones mostly paralleled the pattern of the distribution of geographical area among respective zones. Nearly two-third of the State's net cropped area was concentrated in the zones VI and VII whereas, a little over one-third of the State's net cropped area was dispersed in zones I, II, III, IV, V, VIII and IX.

The critical examination of the land use patterns of the individual agro-climatic zones revealed that during the period, the proportion of the geographical area not used for either forest or cultivation purposes was relatively high in zones I, II and IX. The proportion of the net cropped area to total geographical area was the highest in zone VII followed by zones VIII, VI and V. The share of net cropped area in the total geographical area ranged between about 25 and 50 per cent among zones I, II, III, IV and IX. Moreover, the factors such as undulating land topography, excessive soil losses resulting from high rainfall coupled with water run-off, non-suitability of lands and limitations on the availability of irrigation water resources are some of the prominent reasons for limiting cropped area in these zones.

3.2.3 Cropping Pattern

The State of Maharashtra presents varied topographical situations ranging from hilly areas in Konkan to plain plateau towards Deccan. A greater part of the State is drained to the Bay of Bengal by two of its major rivers, the Godavari and the Krishna, and the minor ones, Wainganga and Penganga, and the rest is drained to the Arabian Sea by west flowing Tapi river and other minor rivers and streams in Konkan. The soils are of varying depth and texture, which are of seven types in the State. The entire State broadly falls under tropical monsoonic climate. However, its major portion is semi-arid with three distinct seasons viz., rainy (June to October), winter (November to February) and summer (March to May). There is considerable variation among different parts of the State. Heavy rain spills over the ghats and coastal districts with annual average of 2000 mm. The major portion of the State, however, lies in the rain shadow of the ghats, with the rainfall averaging around 600-700 mm and in some areas even less than 500 mm. Thus, the rainfall varies from 500 to 5000 mm in the State with average of about 1000 mm distributed over 60 to 70 days. The Southwest monsoon is the main rainy season during which different parts of the State get the major portion of about 80 per cent of the annual precipitation. The variations in the natural resources and climatic factors have blessed the State to have production of a variety of crops. The distribution of area under different crops in different agro-climatic zones for the year 1980-81 and 2000-01 is presented in Appendix - IV and V, respectively. The change in cropping pattern in different agro-climatic zones over the base year 1980-81 is depicted in Table 3.3. These changes are indicated graphically in Chart -3.

Amongst the food crops, sorghum, pearlmillet, paddy and wheat are the important cereals and pigeon pea, green gram, black gram and chick pea are major pulses. In the oilseeds, groundnut, safflower and sunflower are important oilseed crops and sugarcane and cotton are major cash crops of the State.

Table 3.3 : Percentage Change in Cropping Pattern of Different Agro-climatic Zones of Maharashtra

(Figures in percentages)

Crop	*Agro-climatic zones*									
	I	*II*	*III*	*IV*	*V*	*VI*	*VII*	*VIII*	*IX*	*Maharashtra*
Paddy	11.66	-4.36	-53.18	-33.06	-5.78	82.54	-49.18	242.82	5.61	3.70
Kharif sorghum	—	—	796.36	-85.86	-41.92	-53.27	-15.56	-36.11	-85.80	-35.72
Pearlmillet	—	—	290.91	835.54	66.45	2.24	18.22	-60.55	-100.00	17.34
Rabi sorghum	—	—	77.51	138.98	31.56	-11.22	-10.26	-15.05	-79.98	-8.99
Wheat	—	-97.50	-4.51	-27.11	-19.09	-22.81	-44.50	3.21	-51.72	-29.24
Other cereals	-47.92	140.00	-51.96	1823.3	127.11	286.59	7.16	-92.09	-70.32	47.19
Total cereals	-6.15	3.24	-6.29	50.12	18.37	-9.01	-14.72	-25.54	-30.87	-10.51
pigeon pea	—	-2.50	55.21	—	2616.5	1.62	53.54	253.35	51.95	70.20
Chick pea	—	0.00	-54.90	-62.93	-11.98	-1.96	138.87	409.25	-90.60	74.65
Total pulses	104.17	78.33	35.53	67.51	37.29	-15.51	43.30	194.37	-43.72	31.02
Groundnut	—	—	226.08	-6.96	-29.52	-24.53	-56.80	-22.00	-73.06	-29.48
Cotton	—	—	—	—	107.69	-16.16	13.61	51.16	132.30	20.67
Sugarcane	—	—	—	64.13	59.72	52.92	134.44	389.31	132.86	86.55
Fruits & vegetables	797.06	—	874.28	247.78	260.22	377.51	403.36	802.87	544.73	465.85
Gross cropped area	31.45	-4.90	-3.24	11.97	14.18	-7.43	17.23	49.68	-29.38	8.28

Source : Season and crop report, 2002-03, Commissioner of Agriculture, Maharashtra.)

Chart-3 : Changes in Croping Pattern of Maharashtra State during the Period 1980-81 and 2000-01

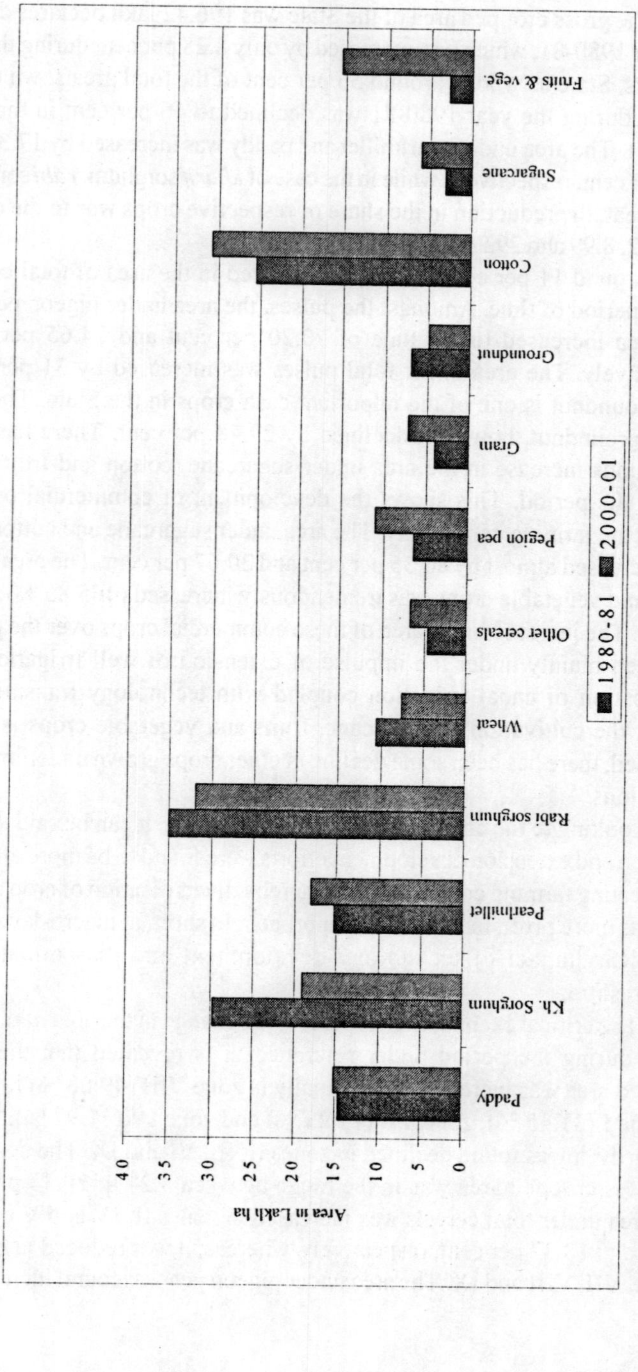

Source: Season and crop report, 2002-03, Commissioner of Agriculture, Maharashtra.

The gross cropped area of the State was 196.42 lakh hectares during the year 1980-81, which was increased by only 8.28 per cent during the last 20 years. State as whole, around 56 per cent of the total area sown under cereals during the year 1980-81 was declined to 46 per cent in the year 2000-01. The area under pearlmillet and paddy was increased by 17.34 and 3.70 per cent, respectively, while in the case of *kharif* sorghum, *rabi* sorghum and wheat, the reduction in the share of respective crops was to the extent of 35.72, 8.99 and 29.24 per cent.

Around 11 per cent decline was noticed in the area of total cereals over a period of time. Amongst the pulses, the area under pigeon pea and chickpea increased to the tune of 70.20 per cent and 74.65 per cent, respectively. The area under total pulses was increased by 31 per cent. The groundnut is one of the important cash crops in the State. The area under groundnut, however, declined by 29.48 per cent. There has been continuous increase in the area under sugarcane, cotton and fruit crops during the period. This shows the development of commercial outlook among the farming community. The area under sugarcane and cotton crop was increased almost by 86.55 per cent and 20.67 per cent. The area under fruits and vegetable crops was tremendously increased (465.85 %) over a period. The increase in the area of these commercial crops over the period has been mainly under the impulse of extension of well irrigation and reallocation of canal irrigation coupled with technology transmission. While, the cultivation of sugarcane, fruits and vegetable crops is being extended, there has been some decline in other crops grown under irrigated conditions.

Looking to the commercialization of cropping, it can be said that the research and extension development efforts were found to be more effective in educating farming community and thereby diversification of crop pattern towards more profit maximization approach. In short, at macro-level there was clear impact of technology development and dissemination in Maharashtra.

The critical examination of percentage change in the area of different crops during the period under reference, it is revealed that the gross cropped area was increased substantially in zone VIII (49.68 %) followed by zone I (31.45 %), zone VII (17.23 %) and zone IV (11.97 %). On the contrarily, it was found declined in zones II, III, VI and IX. The decline in the gross cropped area was in the range between 3.24 to 29.38 per cent. The area under total cereals was increased in zones II, IV and V by 3.24, 50.12 and 18.37 per cent, respectively, whereas, it was reduced in zones I, III, VI, VII, VIII and IX. The area under pigeon pea was found increased in all the zones, excepting zone II. The groundnut area was increased only in zone III, while chickpea area was found increased in zones VII and VIII. The area under cotton, sugarcane, and fruits and vegetable crops was increased substantially in all the zones.

The area share of individual crops in gross cropped area in different agro-climatic zones of Maharashtra is indicated in Table 3.4 (A) and Table 3.4 (B), alongwith the pie Chart- 4.

Chart - 4 : Share of Major Crops in Gross Cropped Area in Maharashtra

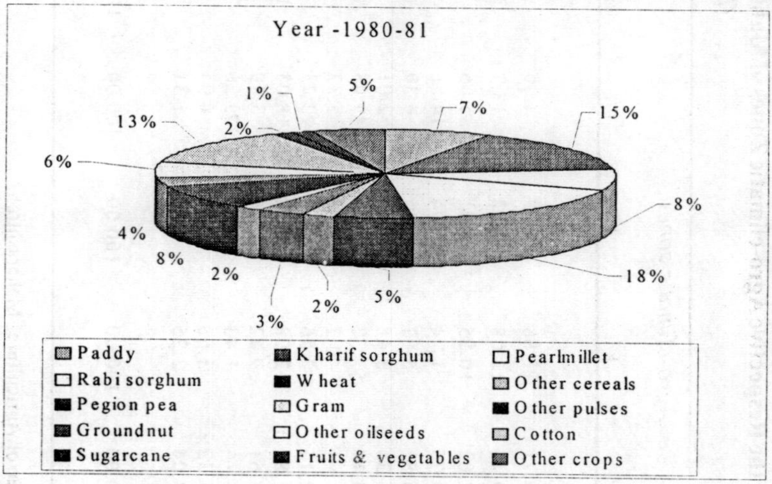

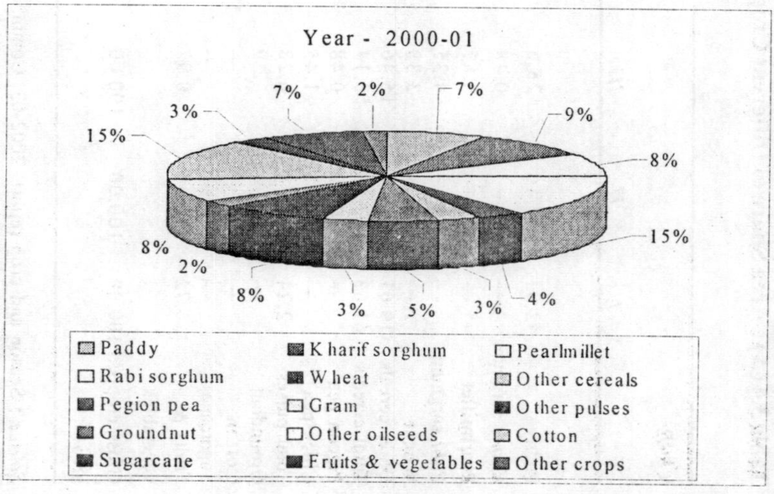

Table 3.4 (A) : Area Share of Different Crops in the Respective Agro-climatic Zones of Maharashtra in 1980-81

(Figures in percentages)

Crop	*Agro-climatic zones*									
	I	*II*	*III*	*IV*	*V*	*VI*	*VII*	*VIII*	*IX*	*Maharashtra*
Paddy	42.02	57.46	25.6	16.71	7.88	0.99	2.06	1.6	37.3	7.42
Kharif sorghum	—	—	0.48	9.49	12.08	9.92	18.7	34.62	8.44	15.13
Pearlmillet	—	—	3.87	2.02	10.85	15.01	5.06	1.2	1.49	7.81
Rabi sorghum	—	—	4.35	9.75	17.4	29.94	14.86	3.29	12.84	17.81
Wheat	—	0.79	3.38	5.82	5.63	5.35	6.39	5.23	4.89	5.41
Other cereals	19.61	3.78	15.46	0.38	3.48	1.08	2.01	3.13	1.56	2.23
Total cereals	61.63	62.03	53.14	44.17	57.32	62.29	49.08	49.1	66.52	55.88
Pigeon pea	—	0.79	0.48	—	0.1	2.98	5.32	3.77	1.56	3.28
Chick pea	—	0.2	1.45	2.15	2.56	2.12	2.73	1.2	1.7	2.08
Total pulses	2.24	3.58	6.28	5.82	12.08	14.08	18.03	10.75	13.83	13.82
Groundnut	—	—	2.9	8.61	9.42	3.96	3.49	1.53	0.5	3.54
Cotton	—	—	—	—	1.43	6.12	23.26	28.96	1.28	12.98
Sugarcane	—	—	—	5.32	3.68	2.52	0.94	0.32	0.07	1.62
Fruits & vegetables	6.72	1.59	0.97	3.04	2.25	1.32	1.31	0.72	0.57	1.41
Gross cropped area	100.00	100.00	100.00	100.00	100.00	100.00	100.00	100.00	100.00	100.00

Source : Season and crop report, 2002-03, Commissioner of Agriculture, Maharashtra).

Table 3.4 (B) : Area Share of Different Crops in the Respective Agro-climatic Zones of Maharashtra in 2000-01

(Area in percentages to the gross cropped area)

Crop	Agro-climatic zones									
	I	*II*	*III*	*IV*	*V*	*VI*	*VII*	*VIII*	*IX*	*Maharashtra*
Paddy	35.71	57.82	12.41	9.98	6.50	1.96	0.89	3.68	55.77	7.11
Kharif sorghum	0.22	—	4.48	1.20	6.14	5.01	13.47	14.78	1.70	8.98
Pearlmillet	—	—	15.64	16.91	15.81	16.57	5.10	0.32	—	8.46
Rabi sorghum	0.20	—	7.99	20.79	20.04	28.72	11.38	1.87	3.64	14.97
Wheat	0.11	0.02	3.34	3.79	3.99	4.46	3.03	3.62	3.34	3.54
Other cereals	7.71	9.46	7.64	6.52	6.92	4.51	1.84	0.16	0.66	3.13
Total cereals	44.02	67.38	51.54	59.20	59.40	61.22	35.70	24.43	65.11	46.18
Pigeon pea	0.14	0.82	0.78	0.54	2.43	3.28	6.96	8.90	3.36	5.15
Chick pea	0.07	0.21	0.68	0.71	1.97	2.24	5.56	4.09	0.23	3.36
Total pulses	3.48	6.72	8.81	8.71	14.56	12.85	22.04	21.14	11.05	16.72
Groundnut	2.36	0.15	9.78	7.15	5.81	3.23	1.29	0.79	0.19	2.30
Cotton	—	—	0.46	0.26	2.61	5.55	22.54	29.25	4.20	14.47
Sugarcane	1.59	—	10.05	7.79	5.15	4.16	1.87	1.05	0.23	2.80
Fruits & vegetables	45.91	24.79	9.74	9.43	7.10	6.82	5.63	4.36	5.18	7.34
Gross cropped area	100.00	100.00	100.00	100.00	100.00	100.00	100.00	100.00	100.00	100.00

Source- Agricultural Statistical Information, Maharashtra State, 2002, Part - II.

It is revealed from these tables that during the year 1980-81, the area under total cereals for State as whole was 55.80 per cent. In different agro-climatic zones, the area under total cereals ranged between 44.17 to 66.52 per cent of the gross cropped area, the lowest and the highest being in zone IV and IX, respectively. In the year 2000-01, the proportion of total cereals in gross cropped area, at the State as whole, was 46.18 per cent. The lowest and the highest share of total cereals was in zone VIII (24.43 %) and zone II (67.38 %), respectively. Over the period the area share of the total cereals was declined. It was considerably reduced in zone I, VII and VIII, whereas, the same was increased in zones II, IV, and V.

In the State, the area share of total pulses in the gross cropped area in the year 1980-81 and 2000-01 was 13.82 per cent and 16.72 per cent, respectively. The proportion area under pulses over the period was found to be increased in all the agro-climatic zones, excepting zones VI and IX. The proportion of area under cotton sugarcane and, fruits and vegetable crops in the gross cropped area indicated increasing trend during the period under study. With regards to the crops, in the year 1980-81, the share of paddy was considerably high in zones I, II, III, IV and IX. The *rabi* sorghum was cultivated on residual soil moisture and hence, its area share in the gross cropped area was concentrated in the zones VI, V, VII and IX, depending upon the types of soils and total annual precipitation, the *kharif* sorghum registered for top in zones VII and VIII, where, *kharif* rains are more and adequately distributed. The area share of wheat was ranged between 0.79 to 6.29 per cent among the zones, excepting zone I. In the year 2000-01, paddy was major in zones I, II and IX. pearlmillet got maximum share in zone III only, *kharif* sorghum covered major share in zones VII and VIII, and *rabi* sorghum accounted higher share in zones VI, V and VI.

The pigeon pea and chickpea are the two major pulses cultivated on large area in the State. The area share of former was ranged between 0.10 to 5.32 per cent in the year 1980-81 and 0.14 to 8.90 per cent in 2000-01, whereas, that of later was between 0.20 to 2.56 per cent in 1980-81 and 0.07 to 5.56 per cent in 2000-01.

During the year 1980-81, there was no cultivation of groundnut, *kharif* sorghum, pearlmillet and *rabi* sorghum in zones I and II, cotton in zones I, II, III and IV, sugarcane in zones I, II and III, and wheat, pigeon pea and chick pea in zone I. In the year 2000-01, there was no cultivation of pearlmillet in zones I, II and IX, *kharif* sorghum in zone II, cotton in zone I and II, and sugarcane in zone II. This indicated that, there existed a diversification in the cropping pattern towards the cash crop during the period under study.

From the proportion of area under individual crops in respective agro-climatic zones, it can be revealed that in zones I and II, monoculturewas predominant, whereas, in the remaining zones, the cropping was diversified in nature. There were wide variations in the cropwise area in the State.

Among the agro-climatic zones also, the change in cropping pattern was noticed and cropping pattern was well diversified in the year 2000-01. The diversification was towards the high value crops with an object to maximise returns, which was aim of research and extension development efforts dealt during last two decades under various programmes. It means among the zones, the impact of technology was also noticed.

3.2.4 Sourcewise Area Under Irrigation

Wells and canals formed the major sources of irrigation in the State. There has been an increase in the gross irrigated area. The break-up of the gross irrigated area according to source of irrigation in different agro-climatic zones of Maharashtra is shown in Table 3.5. The involvement of well irrigation and surface irrigation in total irrigation is indicated in Chart- 5.

Chart-5 : Changes in Irrigation area in Different Agro-Climatie Zones During 1980-81 to 2000-01

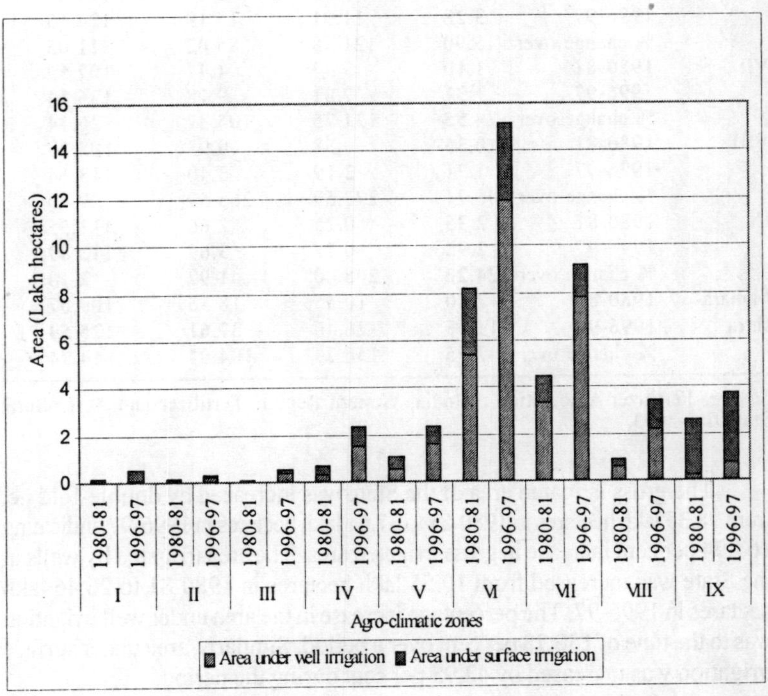

Table 3.5: Sourcewise Irrigated Area in Different Agro-Climatic Zones of Maharashtra

(Area in Lakh hectare)

Zone	Year	Area under surface irrigation	Area under well irrigation	Gross area irrigated	Cropping intensity (%)
I	1980-81	0.13	0.02	0.15	106.25
	1996-97	0.43	0.10	0.53	113.06
	% change over	230.77	385.00	251.33	6.41
II	1980-81	0.07	0.07	0.14	112.78
	1996-97	0.11	0.22	0.33	113.93
	% change over	57.14	214.29	135.71	1.02
III	1980-81	0.05	0.04	0.09	102.99
	1996-97	0.20	0.36	0.56	137.91
	% change over	300.00	800.00	522.22	33.91
IV	1980-81	0.41	0.31	0.72	109.87
	1996-97	0.87	1.48	2.35	131.36
	% change over	112.20	377.42	226.39	19.55
V	1980-81	0.51	0.53	1.04	111.15
	1996-97	0.76	1.62	2.38	123.98
	% change over	49.02	205.66	128.85	11.54
VI	1980-81	2.83	5.38	8.21	111.80
	1996-97	3.28	11.91	15.19	124.15
	% change over	15.90	121.38	85.02	11.05
VII	1980-81	1.10	3.37	4.47	107.59
	1996-97	1.37	7.81	9.18	136.14
	% change over	24.55	131.75	105.37	26.54
VIII	1980-81	0.35	0.58	0.93	103.23
	1996-97	1.21	2.19	3.40	113.14
	% change over	245.71	277.59	265.59	9.60
IX	1980-81	2.35	0.25	2.60	111.55
	1996-97	2.92	0.77	3.69	113.89
	% change over	24.26	208.00	41.92	2.10
Maharashtra	1980-81	7.80	10.55	18.35	109.07
	1996-97	11.15	26.46	37.61	125.54
	% change over	42.95	150.78	104.94	14.94

Source: Fertilizer Association of India, Western Region, Fertilizer and Agricultural Statistics-2003.

The gross irrigated area of the State was increased by double-fold i.e. from 18.35 lakh hectares in 1980-81 to 37.61 lakh hectares in 1996-97, indicating 104.94 per cent increase in gross irrigated area. The area irrigated by wells in the State was increased from 10.55 lakh hectares in 1980-81 to 26.46 lakh hectares in 1996-97. The percentage increase in the area under well irrigation was to the tune of 150.78 per cent over a period. Similarly, area under surface irrigation was increased by 42.95 per cent during the period.

Amongst the agro-climatic zones, the rate of increase in the gross irrigated area was observed higher in zone III (522.22 %) followed by zone VIII (265.59 %) and zone I (251.33 %). The proportion of increase in the gross irrigated area was the lower in zone VI, i.e. only 85.02 per cent over the period. As regards the area under well irrigation, it was observed that the area under well irrigation has increased substantially by 800 per cent in zone III, followed by zone I (385.00 %) and zone IV (377.42 %). The proportion of increase in area under well irrigation was ranged between 121.38 to 800.00 per cent. In case of area under surface irrigation, it was found that the percentage increase in the area under surface irrigation area was higher in zone III (300.00 %) followed by zone VIII (245.71 %) and zone I (230.77 %). The same was found lower in zone VI, i.e. 15.90 per cent only. Moreover, the availability of irrigation water was relatively quite high in zones VI and VII through both the sources. More than half of the State's gross irrigated area seems to have been concentrated in these two zones. The availability of irrigation water was limited in zones I, II and III.

It could also be evidenced from the table that the relative contribution of surface irrigation to the total irrigated area was relatively more than in the zones having higher rainfall as compared to the low rainfall receiving zones.

The availability of irrigation water resource had significant impact on the cropping intensity of the individual zones. For the State as a whole, the cropping intensity has increased from 109.07 per cent in 1980-81 to 125.54 per cent in 1996-97. During the year 1980-81, the cropping intensity varied between 103 and 113 per cent among the zones, the lowest and highest cropping intensity being in zones III and II, respectively. It is quite interesting to note that zone II has got the lowest proportion of its gross cropped area irrigated, but the highest cropping intensity among all the zones. Moreover, the practice of growing short duration crops on residual soil moisture during *rabi* season after the harvest of *kharif* crops seems to have become popular among the farmers in the zone. However, in the year 1996-97, the cropping intensity was found higher in zone III (137.36 %) followed by zone VII (136.14 %) and zone IV (131.36 %). The cropping intensity was in the range between 113.06 per cent and 137.91 per cent.

3.2.5 Livestock Population in Maharashtra

Livestock production forms an important activity next to crop production in the rural areas of the State. The details of livestock population in the year 1980-81 and 1996-97 are given in Table 3.6. The percentage increase in the livestock during the study period is graphically shown in Chart- 6

Table 3.6 : Changes in Livestock Population in Different Agro-climatic Zones of Maharashtra

(Fig. in '000')

Zone	*Year*	*Cattle*	*Buffaloes*	*Sheep*	*Goat & other*	*Total livestock*	*Poultry*
I	1980-81	715.97	140.99	0.80	101.70	959.47	390.92
	1996-97	825.42	250.33	14.75	114.42	2030.33	2567.67
	% change	15.29	77.55	1740.79	12.50	111.61	556.82
II	1980-81	669.10	171.97	4.54	212.65	1058.26	549.67
	1996-97	985.00	402.00	3.00	450.00	2825.00	3545.00
	% change	47.21	133.76	-33.93	111.61	166.95	544.93
III	1980-81	184.24	83.80	7.21	51.62	326.88	861.22
	1996-97	142.65	160.68	71.20	118.54	635.72	662.29
	% change	-22.58	91.74	887.30	129.63	94.48	-23.10
IV	1980-81	804.86	463.89	105.24	337.47	1711.45	2071.69
	1996-97	749.21	474.37	260.34	482.48	2715.61	2812.54
	% change	-6.91	2.26	147.39	42.97	58.67	35.76
V	1980-81	788.70	322.10	179.76	383.70	1674.25	1893.10
	1996-97	704.39	390.73	282.73	581.09	2663.34	2729.13
	% change	-10.69	21.31	57.29	51.45	59.08	44.16
VI	1980-81	5021.47	1363.48	1893.98	3341.54	11620.47	5925.35
	1996-97	5130.94	1765.26	1829.26	4576.63	18433.43	12966.49
	% change	2.18	29.47	-3.42	36.96	58.63	118.83
VII	1980-81	3812.57	928.98	403.05	1981.66	7126.25	4782.35
	1996-97	4901.01	1590.65	602.93	3482.06	15277.66	4489.62
	% change	28.55	71.23	49.59	75.71	114.39	-6.12
VIII	1980-81	2034.77	243.07	29.11	674.93	2981.89	1069.58
	1996-97	2719.95	562.63	80.24	1318.93	7401.90	2371.98
	% change	33.67	131.47	175.61	95.42	148.23	121.77
IX	1980-81	2130.13	253.39	47.28	619.46	3050.26	2299.89
	1996-97	1830.43	365.14	30.14	842.86	4899.00	2334.29
	% change	-14.07		44.10-36.25	36.06	60.61	1.50
Mahara-	1980-81	16161.81	3971.68	2670.96	7704.73	30509.17	19843.79
shtra	1996-------97	17989.00	5961.79	3174.59	11967.01	56881.99	34479.01
	% change	11.31	50.11	18.86	55.32	86.44	73.75

Source : Fertilizer Association of India, Western Region, Fertilizer and Agricultural Statistics-2003.

Chart-6 : Percentage Change in Livestock Population in Differnent Agro-Climatic Zones of Maharashtra During 1980-81 to 2000-01

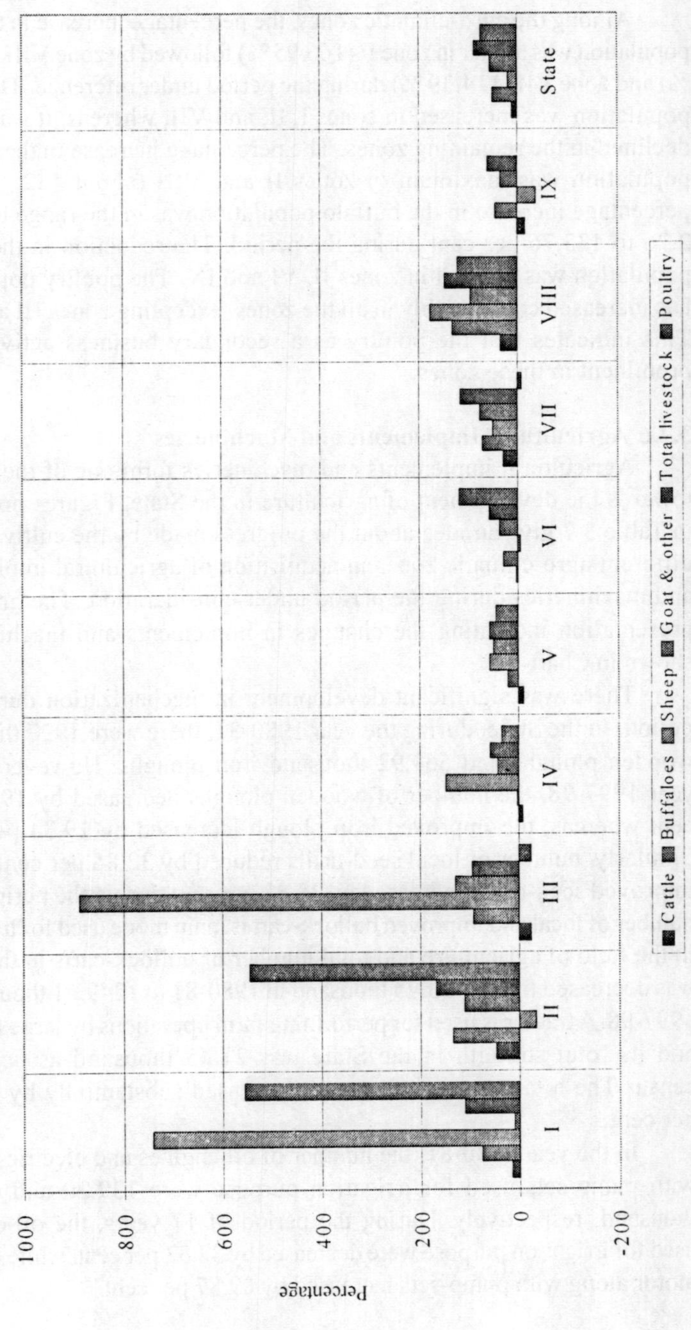

Among the agro-climatic zones, the percentage increase in the total population was higher in zone II (166.95 %) followed by zone VIII (148.23 %) and zone VII (114.39 %) during the period under reference. The cattle population was increased in zones I, II, and VII, whereas, it was found declined in the remaining zones. The percentage increase in the buffalo population was maximum in zones II and VIII (about 132 %). The percentage increase in the buffalo population was in the range between 2.26 to 133.76 per cent during the period. The reduction in the sheep population was noticed in zones II, VI and IX. The poultry population has increased considerably in all the zones, excepting zones III and VII. This indicates that the poultry as a secondary business activity was prominent in these zones.

3.2.6 Agricultural Implements and Machineries

Agricultural implements and machineries form one of the factors towards the development of agriculture in the State. Figures presented in Table 3.7 give an idea about the progress made by the cultivators in different agro-climatic zones in acquisition of agricultural implements and machineries during the period under consideration. The graphical presentation indicating the changes in implements and machinery is shown in Chart- 7.

There was significant development in mechanization during the period. In the State, during the year 1980-81, there were 1927 thousand wooden ploughs and 569.92 thousand iron ploughs. However, in the year 1997-98, the number of wooden ploughs decreased by 19.23 per cent whereas, the improved iron plough increased by 19.34 per cent. Similarly, number of local seed-drills reduced by 32.88 per cent, while improved seed-drills increased by 35.82 per cent during the period. The number of local and improved bullock-cart is main mode used for transport in the field of agriculture and total number of bullock-carts in the State was decreased from 1493.75 thousand in 1980-81 to 1249.61 thousand in 1997-98. A tractor is used for performing farm operations by large farmers and its total strength in the State was 21.45 thousand as per 1982 census.The total number of tractors increased substantially by 279.25 per cent.

In the year 1980-81, the number of oil engines and electric motors with pump-sets used for irrigation purpose were 139.20 and 448.63 thousand, respectively. During the period of 17 years, the oil-engines used for irrigation purpose were decreased by 32.33 per cent, while electric motor along with pump-sets increased by 89.87 per cent.

Table 3.7 : Changes in Agricultural Machinery and Implements in Different Agro-climatic Zones of Maharashtra

(Fig. in '000')

Zone	Year	Plough		Seed-drill		Bullock-carts	Tractors	Oil-engine	Electric-motor
		Wooden	Improved	Local	Improved				
I	1980-81	258.99	1.12	—	—	4.03	0.03	—	—
	1997-98	229.35	2.32	0.83	0.67	6.81	0.50	2.27	7.97
	% change	-11.44	106.80	—	—	68.85	1842.23	—	—
II	1980-81	278.07	1.12	—	—	55.42	0.14	—	—
	1997-98	195.58	2.10	0.21	0.60	47.28	0.44	1.64	3.04
	% change	-29.66	87.19	—	—	-14.69	225.55	—	—
III	1980-81	64.36	2.31	14.84	1.22	19.12	0.17	4.22	1.62
	1997-98	9.50	7.06	6.75	1.38	10.49	1.96	2.13	10.55
	% change	-85.24	205.66	-54.52	12.98	-45.14	1027.93	-49.50	553.22
IV	1980-81	149.73	31.02	32.49	1.69	61.54	3.04	9.31	18.08
	1997-98	50.39	33.01	34.13	3.25	41.22	8.35	8.93	45.97
	% change	-66.35	6.43	5.05	91.88	-33.02	174.29	-4.10	154.26
V	1980-81	130.46	68.83	135.03	6.78	71.70	2.39	8.27	35.62
	1997-98	36.40	72.27	34.07	8.18	37.24	6.95	6.84	48.17
	% change	-72.10	5.00	-74.77	20.59	-48.06	191.07	-17.27	35.23
VI	1980-81	298.30	203.80	474.22	29.31	515.34	11.15	82.04	190.94
	1997-98	228.12	214.94	301.31	35.13	292.81	36.66	34.74	352.79
	% change	-23.53	5.47	-36.46	19.84	-43.18	228.75	-57.66	84.77
VII	1980-81	247.04	215.79	299.21	23.75	316.68	3.73	27.20	120.77
	1997-98	252.92	236.27	376.54	31.27	402.02	18.90	19.25	274.38
	% change	2.38	9.49	25.84	31.65	26.95	406.32	-29.23	127.19

(Contd.)

(Contd. Table)

Zone	Year	Plough		Seed -drill		Bullock-carts	Tractors	Oil-engine	Electric-motor
		Wooden	Improved	Local	Improved				
VIII	1980-81	239.91	37.88	222.20	8.49	209.87	0.37	6.22	72.77
	1997-98	241.01	86.27	125.73	13.37	217.38	4.25	9.65	94.42
	% change	0.46	127.75	-43.42	57.57	3.58	1045.13	55.09	29.75
IX	1980-81	260.15	8.05	158.96	10.18	240.05	0.43	1.93	8.84
	1997-98	313.17	25.93	17.74	16.75	194.36	3.35	8.75	14.53
	% change	20.38	222.07	-88.84	64.56	-19.03	676.89	352.24	64.40
Maharashtra	1980-81	1927.00	569.92	1336.96	81.43	1493.75	21.45	139.20	448.63
	1997-98	1556.44	680.17	897.31	110.60	1249.61	81.36	94.20	851.82
	% change	-19.23	19.34	-32.88	35.82	-16.34	279.25	-32.33	89.87

Source: Agricultural Statistical Information, Maharashtra state, 2002, part-II.

Chart-7 : Changes in implements and machinery in Maharashtra during the period 1980-81 to 1997-98

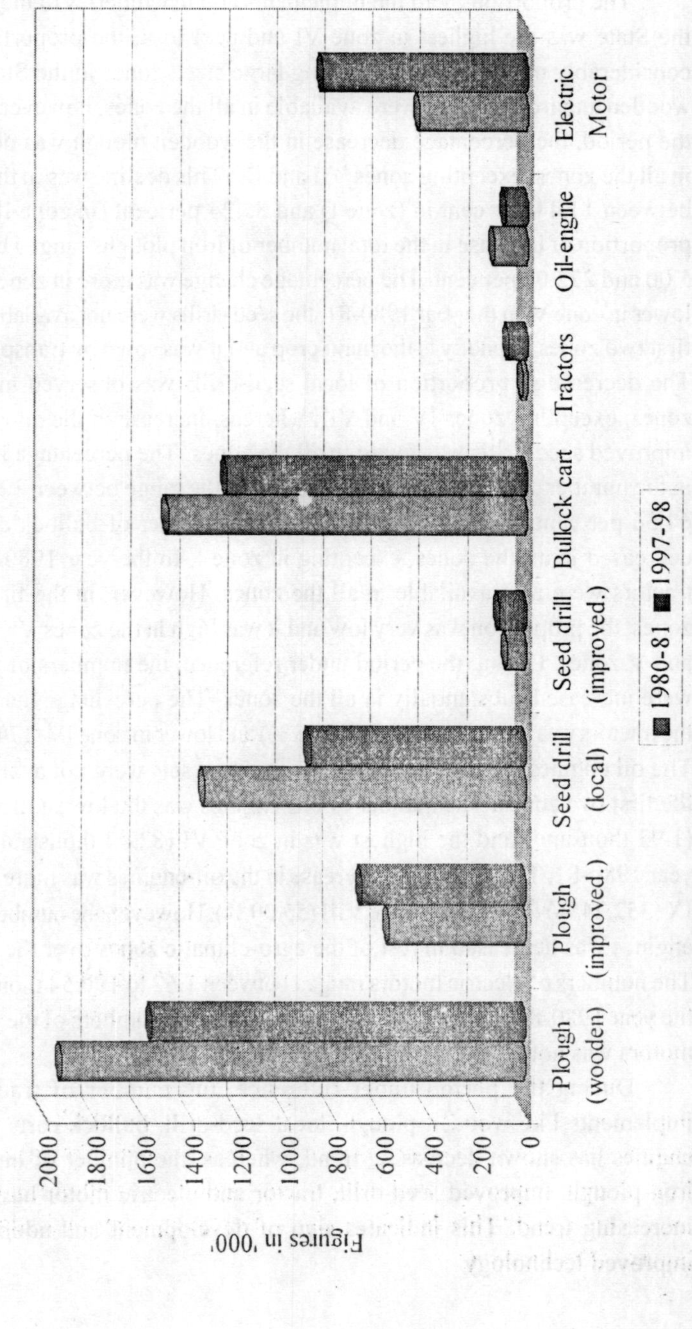

The proportion of all the implements and machineries to the total of the State was the highest in zone VI and next to it, the proportion was considerably more in zone VII, being large sized zones in the State. The wooden and iron ploughs were available in all the zones, however, during the period, the percentage decrease in the wooden plough was observed in all the zones, excepting zones VII and IX. This decline was in the range between 11.44 per cent in (zone I) and 85.24 per cent (in zone III). The proportion of increase in the total number of iron ploughs ranged between 5.00 and 222.07 per cent. The percentage change was more in zone IX and lower in zone V. In the year 1980-81, the seed-drills were not available in the first two zones as paddy is the main crop and it was sown by transplanting. The decrease in proportion of local seed-drills was observed in all the zones, excepting zones IV and VII, whereas, increase in the adoption of improved seed-drills was found in all the zones. The percentage increase in the number of improved seed-drills was in the range between 12.98 and 64.56 per cent during the period. The total number of bullock cart was decreased in all the zones, excepting in zone I. In the year 1980-81, the tractors were also available in all the zones. However, in the first three zones, the proportion was very low and it was high in the zones VI, VII, IV and V zones. During the period under reference, the numbers of tractors were increased substantially in all the zones. The percentage increase in the tractors was more in zone I (1842.23 %) and lower in zone IV (174.29 %). The oil engines and electric motors with pump sets were not available in the first two zones. The number of oil- engines was the lowest in zone IX (1.93 thousand) and the highest was in zone VI (82.04 thousand) in the year 1980-81. The percentage increase in the oil-engines was more in zone IX (352.24 %) followed by zone VIII (55.09 %). However, the number of oil-engines was decreased in rest of the agro-climatic zones over the period. The numbers of electric motors ranged between 1.62 to 190.54 thousand in the year 1980-81. The considerable increase in the numbers of the electric motors was noticed in all the zones during the period.

During the period under reference, the number of traditional implements like wooden plough, local seed-drill, bullock-carts and oil engines has shown decreasing trend, whereas, the number of improved iron plough, improved seed-drill, tractor and electric motor has shown increasing trend. This indicates sign of development and adoption of improved technology.

3.2.7 Productivity Scenario of Major Crops in Maharashtra

It would be appropriate to examine the impact of the developmental and technological efforts on the performance of agriculture in the State. Development in the agricultural activity can be reflected through the improvement in the agricultural productivity. The only quantitative indicator of agricultural productivity is the yield per hectare, about which the information is available. The productivity of major crops according to different agro-climatic zones of Maharashtra during the pre-technological and post-technological period is depicted in Table 3.8, and graphically presented in Chart- 8.

In the state as a whole, during the period of 20 years, there was a considerable improvement in the productivity of all the crops, except sugarcane. The increase in the productivity of these crops was in the range between 10.65 to 110.42 per cent. The improvement in yields of paddy, *kharif* sorghum, pearlmillet, pigeon pea, *rabi* sorghum, wheat, chick pea, and cotton could be attributed to the adoption of improved varieties and the improved agronomic practices.

The performance of sugarcane was observed to be disappointing as the per hectare yield of this crop declined by about 9 per cent during the period. This indicated that the present crop production and management technology of this crop is remained stagnant. There is a need to create awareness among the farmers for using the improved varieties and adoption

The scenario of the productivity of major crops is given in Table 3.9. The actual yield in Maharashtra is compared graphically with potential yield and national average and shown in Chart- 9.

It can be seen from the Table 3.9 that, the productivity of selected crops in Maharashtra, was substantially lower by 24.16 to 83.27 per cent than that of potential yields of these crops in the situation of recommended use of inputs. It clearly indicated that, the crop production technology was not fully exploited during the period under study. There existed a large scope to increase the productivity of these crops by adopting the crop production and management technology at its extreme. In the analysis of technology adoption level, the magnitude of composite index of technology adoption has indicated the low level of technology adoption in the production of crops.

The low adoption of the agricultural technology was the probable reason for low productivity of these crops. Besides, the agro-ecological situations in the State, scanty and irregular rainfall, large coverage of drought-prone area, etc. are the other reasons that hampered the rainfed crop productivity. In case of sugarcane, the state productivity touched to

Table 3.8 : Productivity of Major Crops in Different Agro-climatic Zones of Maharashtra

(Figures in qtl / ha)

Zone	Year	Paddy	Kh. sorghum	Pearlmillet	Pigeon pea	Rabi sorghum	Wheat	Chick pea	Groundnut	Cotton *	Sugarcane
I	1980-81	18.09	—	—	—	—	—	—	—	—	—
	2000-01	21.90	—	—	—	—	—	—	—	—	—
	% change	21.03	—	—	—	—	—	—	—	—	—
II	1980-81	18.30	—	—	2.20	—	8.36	3.13	—	—	—
	2000-01	19.90	—	—	3.59	—	10.00	5.34	—	—	—
	% change	8.74	—	—	63.53	—	19.62	71.03	—	—	—
III	1980-81	19.56	13.52	4.67	—	6.07	10.07	4.70	9.11	—	—
	2000-01	18.65	16.29	5.51	—	6.18	15.20	5.55	13.07	—	—
	% change	-4.68	20.49	18.00	—	1.81	50.95	18.05	43.45	—	—
IV	1980-81	19.66	14.62	4.96	7.78	4.58	9.42	4.32	8.95	—	914.20
	2000-01	17.46	12.66	5.83	4.10	4.46	11.85	5.86	12.57	—	916.30
	% change	-11.19	-13.39	17.51	-47.32	-2.57	25.73	35.83	40.36	—	0.23
V	1980-81	16.70	12.62	4.39	6.56	5.00	9.39	3.89	8.31	0.73	949.00
	2000-01	15.60	11.32	5.55	4.06	4.29	12.78	5.33	10.27	0.96	883.10
	% change	-6.59	-10.34	26.38	-38.12	-14.29	36.13	37.23	23.58	31.51	-6.94
VI	1980-81	15.17	10.53	4.43	3.61	4.38	8.95	3.30	6.36	0.82	953.70
	2000-01	11.92	12.08	5.91	4.38	4.71	13.96	5.51	8.50	0.88	853.30
	% change	-21.43	14.79	33.24	21.55	7.63	55.93	66.86	33.73	7.32	-10.53
VII	1980-81	5.70	9.95	4.72	5.03	4.57	8.31	2.95	4.58	0.45	807.80
	2000-01	7.90	13.27	6.83	5.21	6.06	12.26	5.07	4.63	1.06	745.50
	% change	38.60	33.45	44.81	3.58	32.55	47.53	71.98	1.09	135.56	-7.71

(Contd.)

(Table Contd.)

Zone	Year	Paddy	Kh.	Pearlmillet	Pigeon	Rabi	Wheat	Chick	Groundnut	Cotton *	Sugarcane
VIII	1980-81	8.30	7.64	3.81	6.61	4.40	7.28	3.53	4.64	0.42	772.50
	2000-01	10.60	11.68	6.19	8.76	5.01	11.48	4.66	6.32	0.98	682.90
	% change	27.71	52.92	62.71	42.17	13.87	57.72	32.23	36.26	133.33	-11.60
IX	1980-81	8.80	5.82	—	4.23	4.37	5.71	4.86	6.41	0.37	747.90
	2000-01	10.30	14.19	—	5.78	2.64	7.22	4.34	6.27	0.71	714.00
	% change	17.05	143.92	—	36.82	-39.54	26.54	-10.67	-2.21	91.89	-4.53
State	1980-81	15.87	9.56	4.54	4.95	4.48	8.34	3.35	6.10	0.48	917.40
	2000-01	17.56	12.56	6.04	6.02	4.99	12.56	5.19	10.81	1.01	832.67
	% change	10.65	31.34	33.05	21.79	11.38	50.63	54.84	77.31	110.42	-9.24

* Productivity of cotton in bales / ha. and 1 bale = 170 kg

Source: Agricultural Statistical Information , Maharashtra state, 2002, Part- II

Chart 8: Changes in Productivity of Major Crops during the Year 1980-81 and 2000-01

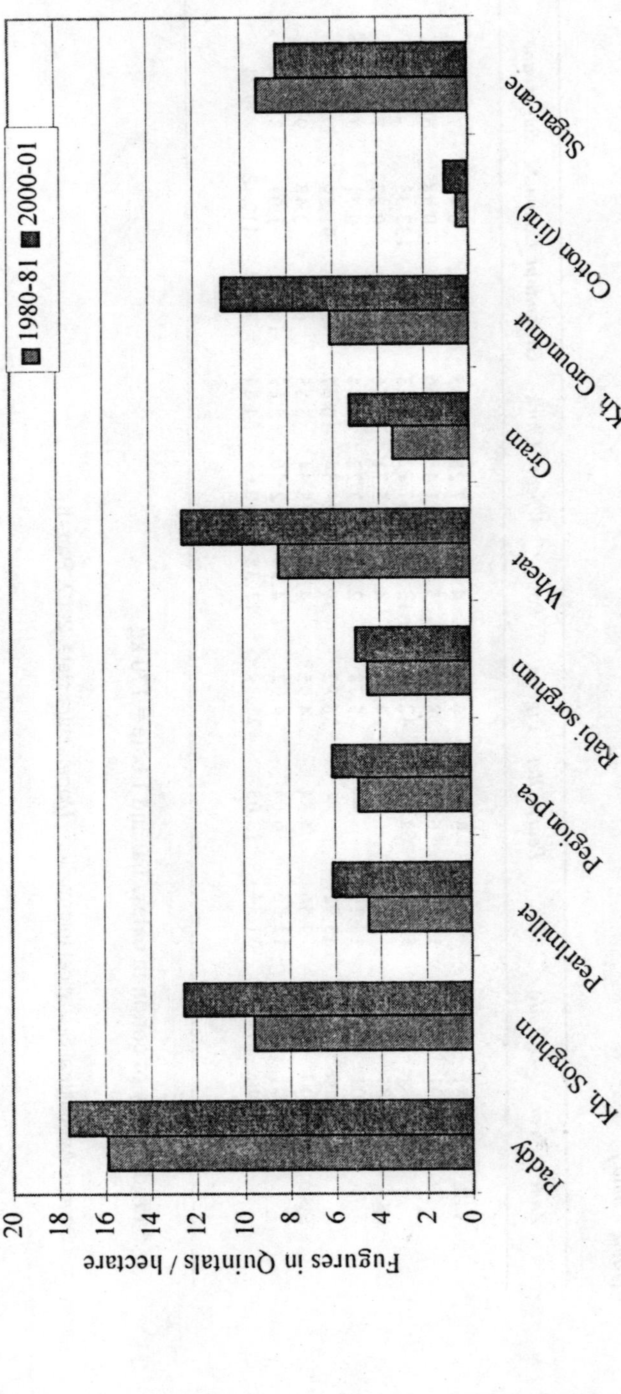

Note- Productivity of sugarcane in {(tonnes/ ha)/10} and cotton in bales /ha. (1 bale = 170 kg.)

Chart 9: Productivity Scenario of Major Crops at State and National level 2000-01

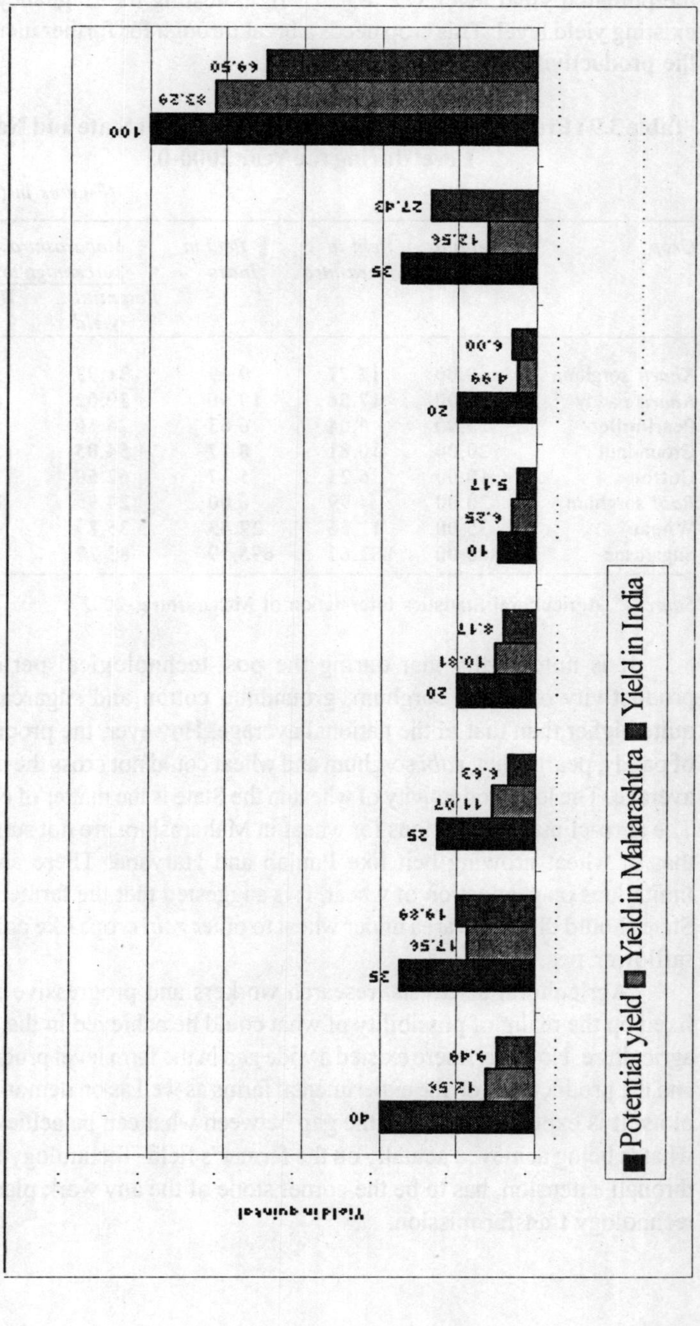

the potential yield level, (i.e. 83.27 %), indicating the stagnancy in the existing yield level. This crop needs a breakthrough for further increase in the productivity.

Table 3.9 : Productivity Scenario of Major Crops at State and National Level during the Year 2000-01

(Figures in (qtl / ha)

Crop	*Potential yield level*	*Yield in Maharashtra*	*Yield in India*	*Maharashtra's yield percentage to*	
				Potential yield	*Yield in India*
Kharif sorghum	40.00	12.77	9.49	31.93	134.51
Kharif paddy	45.00	17.56	19.90	39.02	88.25
Pearlmillet	25.00	6.04	6.63	24.16	91.12
Groundnut	20.00	10.81	8.17	54.05	132.37
Cotton	10.00	6.25	5.17	62.50	120.89
Rabi sorghum	20.00	4.99	6.00	24.95	83.12
Wheat	35.00	12.56	27.43	35.89	45.79
Sugarcane	1000.00	832.67	695.19	83.27	119.78

Source : Agricultural Statistics Information of Maharashtra -2002.

It is noteworthy that during the post-technological period, the productivity of *kharif* sorghum, groundnut, cotton and sugarcane was quite higher than that of the national average. However, the productivity of paddy, pearlmillet, *rabi* sorghum and wheat could not cross the national average. The low productivity of wheat in the State is the matter of concern. The agro-climatic conditions for wheat in Maharashtra are not suitable as that of wheat growing belt like Punjab and Haryana. There are some limitations on production of wheat. It is suggested that the farmers in the State should divert the area under wheat to other *rabi* crops like chick pea, safflower, pea, etc.

Agricultural scientists, research workers and progressive farmers fixed up the realm of possibility of what could be achieved in the field of agriculture. However, there existed a wide gap in the farm level productivity and the productivity on the experimental farms as well as on demonstration plots. It is expected to bridge the gap between what can be achieved and what is being achieved actually on the farmer's field. Technology transfer through extension, has to be the corner stone of the any work plan of the technology transfer mission.

4

Impact of Technology on Resource Use, Cost and Returns of Major Crops

The new technology in agriculture has generated optimism for prosperity both on commercial and subsistence farms. It means that at the micro level, a farm is subjected to change its resource use structure in response to technical advancement, price changes and agricultural infrastructure development in the region. The constant returns of individual crop enterprises are also influenced largely due to the variations in the resource use structure at the farm level in different regions. The purpose of this chapter is, therefore, to analyze and understand the differences in resource use structure, production costs and returns of each crop in different agro-climatic zones of Maharashtra during the period of last twenty years, i.e. from 1980-81 to 2000-01. The analysis contemplated here is expected to be highly useful to provide a basis for understanding the differences in the resource productivities and allocation efficiency of the sample farms across the zones.

4.1 Adoption Level of Technology on Sample Farms

Technological change is one of the major forces leading to changes in output, employment and functional income distribution. Although the technological breakthrough as such in Maharashtra's agriculture is not new phenomenon, yet it has a significant relevance particularly for the regions, which are still lying in the embryonic stage of agricultural development. (Chandra, 1992). The scarcity dominant regions in

Maharashtra falling under such conditions of agricultural backwardness have not yet been able to benefit from the fruits of the technological innovation fully or even partly.

The new agricultural technology is believed to have generated more income to the cultivators. Several studies on technology adoption have separately addressed either factors affecting in adoption or its impact on farm economy across the farms and regions. The studies by Pandy and Murlidharan (1977), Agarwal (1983), Yadav and Gangwar (1983), and Kiresur et al. (1996) revealed that economic prosperity and social mobilization and favourable psychological motivation are instrumental in the adoption of new technology. Sohoni and Khandarkar (1970) have pointed out that in terms of income growth, small farmers have done better than the large ones, while, Frankel (1971) brought out that the large farmers have benefitted more than the small ones from the technology. The benefits accrued through improved technology were estimated across regions or with particular factor. In order to increase adoption level and income of farmers, the NARP approach was initiated since 1982 among agro-climatic zones, wherein research-extension linkage was accelerated. In this regards, Kapase (1992) has attempted to estimate the resource use productivities in different agro-climatic zones of Maharashtra during the year 1980-81. It was concluded that there was differentials in resource use productivities and existed inefficiencies in resource use. A study on impact of improved technology across the agro-climatic zones was not attempted in the country.

A comprehensive study incorporating the interactions between level of adoption and economic gains is necessary to get an integrated view. Thus, an attempt was made to study the extent of adoption of crop production technology at micro level in different agro-climatic zones of Maharashtra. The recommendations involved in crop production are of two categories viz., qualitative and quantitative recommendations. In this study, the quantitative recommendations were considered for determining the impact of agricultural technologies at two points of times i.e. 1980-81 and 2000-01. The technology index was developed on the basis of use levels of five key inputs like variety, seedrate, manure, fertilizers and plant protection measures as per the procedure explained in Chapter 2. The components of recommended crop production technology by the State Agricultural Universities (SAU's) were identified and the same are presented cropwise in Appendix VI. The extent of adoption of recommendation by the farmers in different agro-climatic zones with respect to each component of crop production technology has been quantified and presented in Table 4.1 (A) to 4.1 (H), alongwith the graphical presentation indicated in Chart- 10 to Chart- 17.

***Kharif* sorghum :** The total technology adoption index at the overall level was raised by 42.17 per cent. The adoption of recommended varieties was increased by 439.83 per cent followed by fertilizers. The adoption level of manure was found to be declined. The adoption of high yielding varieties in both the zones under study was substantially increased to the tune of 633 per cent.

Table 4.1 (A) : Technology Adoption Index for *kharif* Sorghum

(Per cent)

Technology	*1980-81*			*2000-01*		
Parameters	*VII*	*VIII*	*Overall*	*VII*	*VIII*	*Overall*
Variety	9.09	4.76	6.98	66.67	34.92	37.68
				(633.44)	(633.61)	(439.83)
Seedrate	62.12	47.62	55.04	48.48	63.49	55.81
				(-21.96)	(33.33)	(1.40)
Manure	28.03	12.70	20.54	12.88	5.56	9.30
				(-54.05)	(-56.[illegible])	(-54.72)
Fertilizers	42.92	5.26	24.42	47.73	[illegible]0.32	53.88
				(11.21)	(1046.77)	(120.64)
Plant protection	9.09	28.27	18.60	0.00	0.00	0.00
				—	—	—
Technology adoption index	33.97	15.54	24.97	34.69	36.34	35.50
				(2.12)	(133.85)	(42.17)

(Figures in the parentheses indicate the percentage change over the base year)

The adoption of recommended level of seedrate was remained by and large, the same (i.e. 55 %) during the period. The application level of farm yard manure was also relatively reduced in both the zones. This was because of less use of manure due to tendency of farmer to use available and limited manure to cash crops preferably. The index for plant protection

Chart-10: Technology Adoption Index for Kharif Sorghum in Different Agro-Climatic Zones

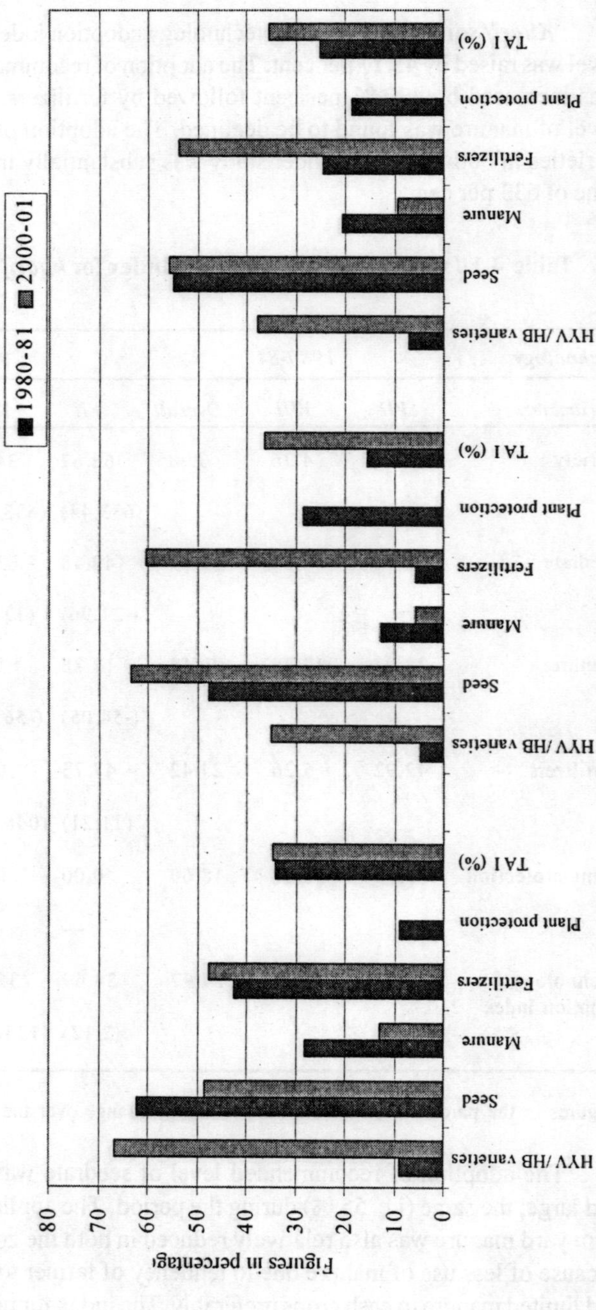

measures was noticed in both the zones in the year 1980-81 only. The considerable increase in total technological index was observed in zone VIII (133.85 %) as compared to zone VII. The technology adoption index was found to be increased during the period of twenty years.

***Kharif* paddy :** It is observed that the farmers were cultivating the local varieties during the year 1980-81 and there was substantial increase in adoption of high yielding varieties during the year 2000-01, and its level of adoption was to the extent of 83.33 per cent in zone V and it was 78.33 and 66.67 per cent in zones IX and II, respectively.

Table 4.1 (B) : Technology Adoption Index for *kharif* Paddy

(Per cent)

Technology	*1980-81*				*2000-01*			
Parameters	*II*	*V*	*IX*	*Overall*	*II*	*V*	*IX*	*Overall*
Variety	0.00	0.00	0.00	0.00	66.67	83.33	78.33	76.27
					—	—	—	—
Seedrate	59.65	56.67	55.00	56.44	61.67	66.67	65.00	64.83
					(3.39)	(17.65)	(18.18)	(14.87)
Manure	6.67	7.50	22.50	12.22	42.98	2.50	5.00	16.38
					(544.38)	(-66.67)	(-77.78)	(34.04)
Fertilizers	7.50	14.17	3.33	8.33	10.53	21.67	46.67	26.55
					(40.40)	(52.93)	(1301.5)	218.73
Plant protection	5.00	000	000	1.67	0.00	0.00	0.00	0.00
					—	—	—	—
Technology adoption index	14.47	17.37	18.42	16.75	36.84	29.74	38.95	35.15
					(154.60)	(71.21)	(111.45)	(109.85)

(Figures in the parentheses indicate the percentage change over the base year)

The level of adoption of recommended seedrate was increased by about 18 per cent in zones V and IX, while in zone II it was increased by only 3.39 per cent. The adoption for manure was reduced by 67 to 78 per cent in zones V and IX, whereas it was substantially increased in zone II (544.48 %). The adoption level of fertilizers was found to be increased to

Chart-11: Technology Adoption Index for Kharif Paddy in Different Agro-climatic Zones

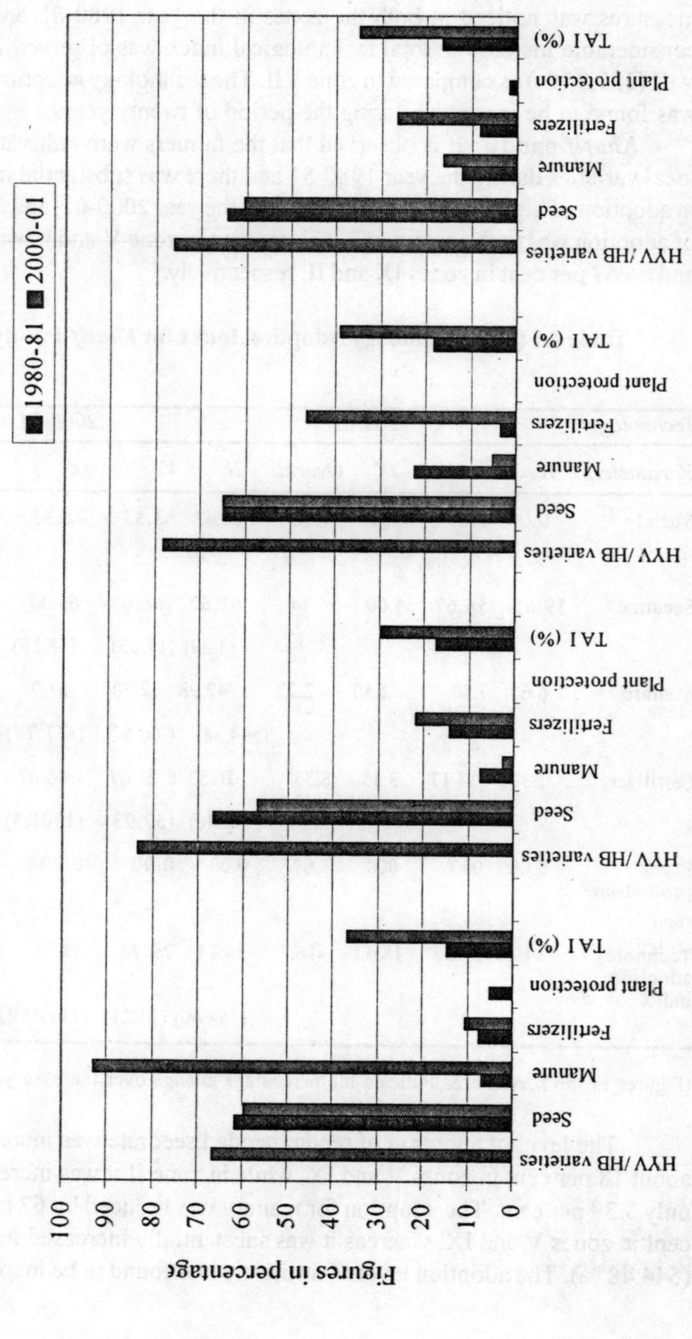

large extent in zone IX followed by zones V and II. At the overall level, its adoption was increased by 218.73 per cent. The total technology index was increased from 16.75 per cent in 1980-81 to 35.15 per cent in 2000-01, at the overall level, while, it was increased to large extent in zones II followed by zones IX and V. This means the technology adoption was found to be increased during the period under study.

Pearlmillet : The adoption levels of varieties, seedrate, manure and fertilizer resource inputs were found to be increased among the zones, excepting manure in zone VII, wherein no use was noticed.

Table 4.1 (C) : Technology Adoption Index for Pearlmillet

(Per cent)

Technology	*1980-81*				*2000-01*			
Parameters	*V*	*VI*	*VII*	*Overall*	*V*	*VI*	*VII*	*Overall*
Variet	6.67	1.67	0.00	3.09	43.33	36.67	57.14	44.44
					(549.63)	(2095.8)	—	(1338.2)
Seedrate	43.33	36.67	35.71	38.89	73.33	65.00	66.67	68.52
					(69.24)	(77.26)	(86.70)	(76.19)
Manure	4.17	4.17	5.95	4.63	8.33	20.83	0.00	10.80
					(99.76)	(399.52)	—	(133.26)
Fertilizers	0.00	5.83	4.76	3.40	35.83	26.67	34.52	32.10
					—	(357.46)	(625.21)	(844.12)
Plant protection	0.00	0.00	0.00	0.00	15.00	0.00	0.00	5.56
					—	—	—	—
Technology index	9.21	9.21	9.02	9.16	33.66	31.05	30.45	31.68
					(265.47)	(237.13)	(237.58)	(245.85)

(Figures in the parentheses indicate the percentage change over the base year)

It was also noticed that, no plant protection care was undertaken during both the period. The increase in adoption level was found to be the highest in varieties due to use of high yielding varieties of pearlmillet, followed by fertilizers, manure and seedrate. The total technology adoption

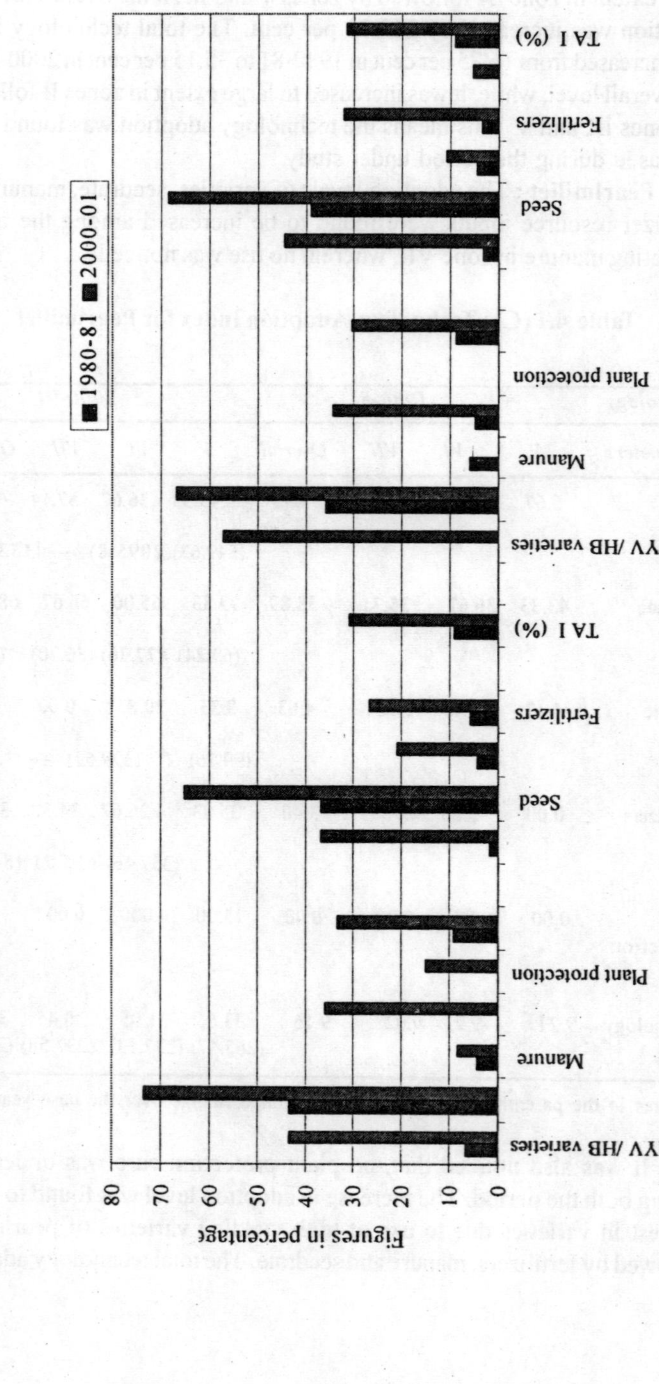

Chart-12: Technology adoption index for pearlmillet in different agro-climatic zones

index was increased among all the zones and the extent of increase in adoption was 245.85 per cent at the overall level.

Groundnut : The coverage of high yielding varieties on sample farm was increased by 162.47 and 200.05 per cent in zones V and VII, respectively. Similarly, score for fertilizers was increased by 19.06 per cent in zone V and 128.76 per cent in zone VII. The adoption level for seedrate and manure was reduced by 36.10 and 42.11 per cent, at the overall level.

Table 4.1 (D) : Technology Adoption Index for Groundnut

(Per cent)

Technology	*1980-81*			*2000-01*		
Parameters	*V*	*VI*	*Overall*	*V*	*VI*	*Overall*
Variety	26.67	22.22	24.39	70.00	66.67	68.33
				(162.47)	(200.05)	(180.16)
Seedrate	48.33	76.19	62.60	45.00	35.00	40.00
				(-6.89)	(-54.06)	(-36.10)
Manure	50.00	19.84	34.55	30.83	9.17	20.00
				(-38.34)	(-53.78)	(-42.11)
Fertilizers	35.00	22.22	28.46	41.67	50.83	46.25
				(19.06)	(128.76)	(62.51)
Plant protection	0.00	0.00	0.00	10.00	0.00	5.00
				—	—	—
Technology adoption index	38.68	28.82	33.63	41.58	35.00	38.29
				(7.50)	(21.44)	(13.86)

(Figures in the parentheses indicate the percentage change over the base year)

The adoption of seedrate was decreased due to high price of groundnut seeds, whereas, due to less availability of manure, the adoption of manure was reduced. The technology adoption index was increased by only 13.86 per cent during the period, at the overall level, while it was increased more in zone VII as compared to zone V.

Chart-13: Technology adoption index for groundnut in different agro-climatic zones

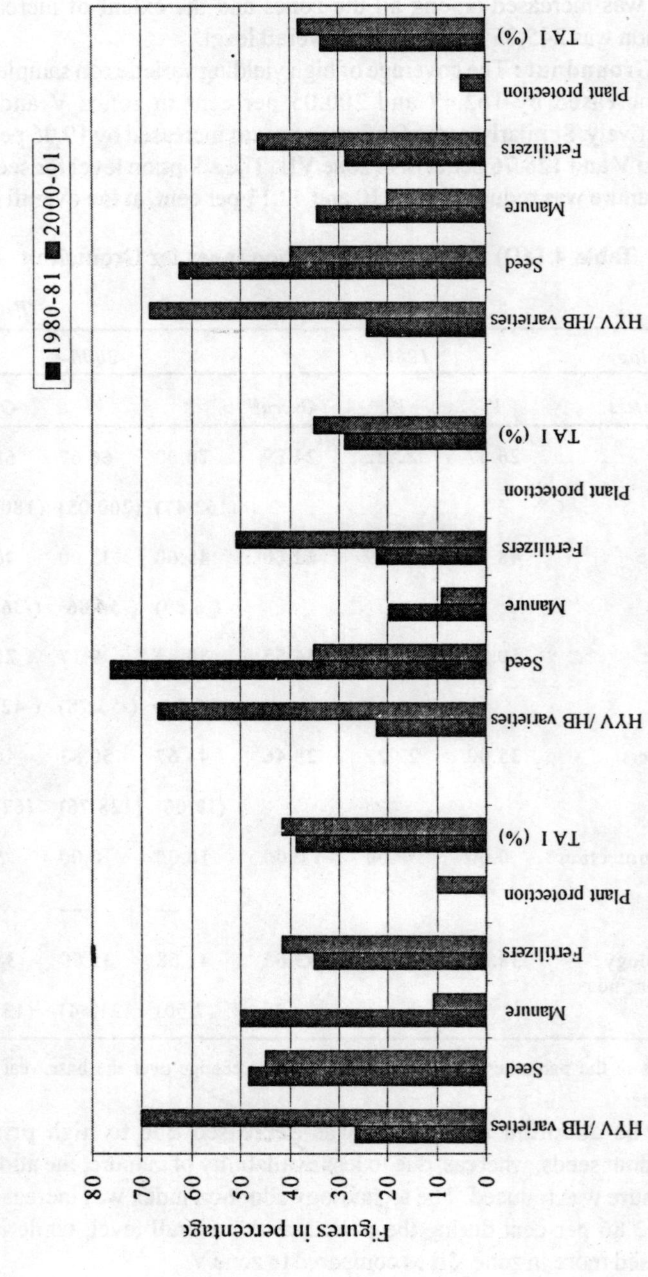

Cotton : The adoption level was found to be increased in case of varieties, manure and fertilizers, at the overall level. Similar was the situation on the cotton farms in zones VII and VIII, excepting manure in zone VIII. In case of seedrate, the adoption score was found to be increased in zone VIII, while reverse was the trend in zone VII and at the overall level. The plant protection care was increased in zone VII, while, the expenditure on it, was decreased in zone VIII and at the overall level.

Table 4.1 (E) : Technology Adoption Index for Cotton

(Per cent)

Technology Parameters	*1980-81*			*2000-01*		
	VII	*VIII*	*Overall*	*VII*	*VIII*	*Overall*
Variety	40.35	48.33	44.44	45.61	55.00	50.43
				(13.04)	(13.80)	(13.48)
Seedrate	61.40	50.00	55.56	43.86	65.00	54.70
				(-28.57)	(30.00)	(-1.55)
Manure	7.02	19.17	13.25	17.54	11.67	14.53
				(149.86)	(-39.12)	(9.66)
Fertilizers	27.19	20.00	23.50	47.37	55.83	51.71
				(74.22)	(179.15)	(120.04)
Plant protection	52.63	55.00	53.85	57.89	25.00	41.03
				(9.99)	(-54.55)	(-23.81)
Technology adoption index	29.64	30.79	30.23	37.67	41.58	39.68
				(27.09)	(35.04)	(31.26)

(Figures in the parentheses indicate the percentage change over the base year)

***Rabi* sorghum :** The adoption of high yielding varieties and plant protection measures was noticed in 2000-01 to the extent of 33.33 per cent and 2.78 per cent, respectively, at the overall level. The use of manure was reduced by 31.97 per cent in zone VI and 86.94 per cent in zone VII over a period. This was due to reduction in availability of manure and tendency of farmers to use this available manure for cash crops. The adoption of

Chart-14: Technology Adoption Index for Cotton in Different Agro-climatic Zones

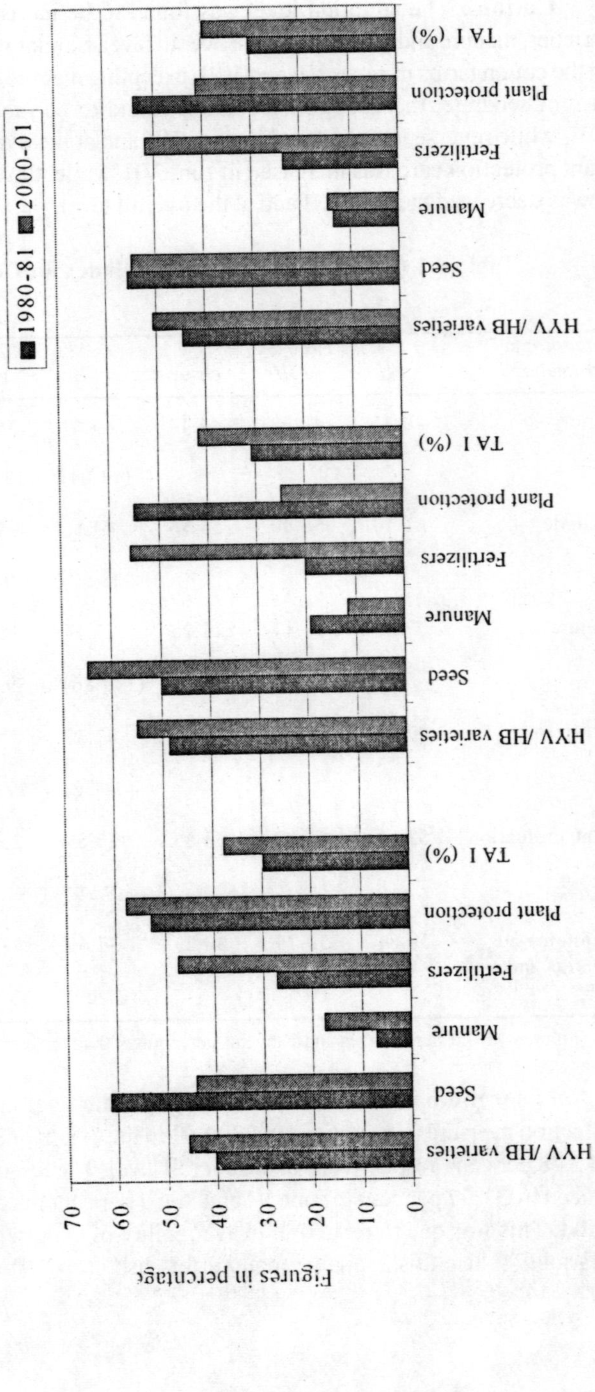

recommended level of seedrate was increased among the zones and at the overall level. It was increased from 60.32 per cent to 78.70 per cent during the period. The technology index was increased by 32.08 per cent in zone VI and 13.71 per cent in zone VII.

Table 4.1 (F) : Technology Adoption Index for *rabi* Sorghum

(Per cent)

Technology	*1980-81*			*2000-01*		
Parameters	*VI*	*VII*	*Overall*	*VI*	*VII*	*Overall*
Variety	0.00	0.00	0.00	33.33	33.33	33.33
				—	—	—
Seedrate	68.33	58.33	63.89	83.33	72.92	78.70
				(21.95)	(25.01)	(23.18)
Manure	20.83	23.96	22.22	14.17	3.13	9.26
				(-31.97)	(-86.94)	(-58.33)
Fertilizers	12.50	0.00	6.94	15.83	4.17	10.65
				(26.64)	—	(53.46)
Plant protection	0.00	0.00	0.00	5.00	0.00	2.78
				—	—	—
Technology adoption index	21.32	16.78	19.30	28.16	19.08	24.12
				(32.08)	(13.71)	(24.97)

(Figures in the parentheses indicate the percentage change over the base year)

Wheat : The coverage of high yielding varieties and use of fertilizers were substantially increased in all the zones under study. This was due to availability of quality seeds of high yielding as well as fertilizer responding varieties in the market. The adoption index for seedrate was reduced in all the zones. This was due to availability of quality seeds as well as the availability of improved seed-drills of suitable distance as per the recommendations. The total technology adoption index was increased in all the three zones, in the range between 17.51 per cent to 33.01 per cent during the period.

Chart-15: Technology adoption index for rabi sorghun in different agro-climatic zones

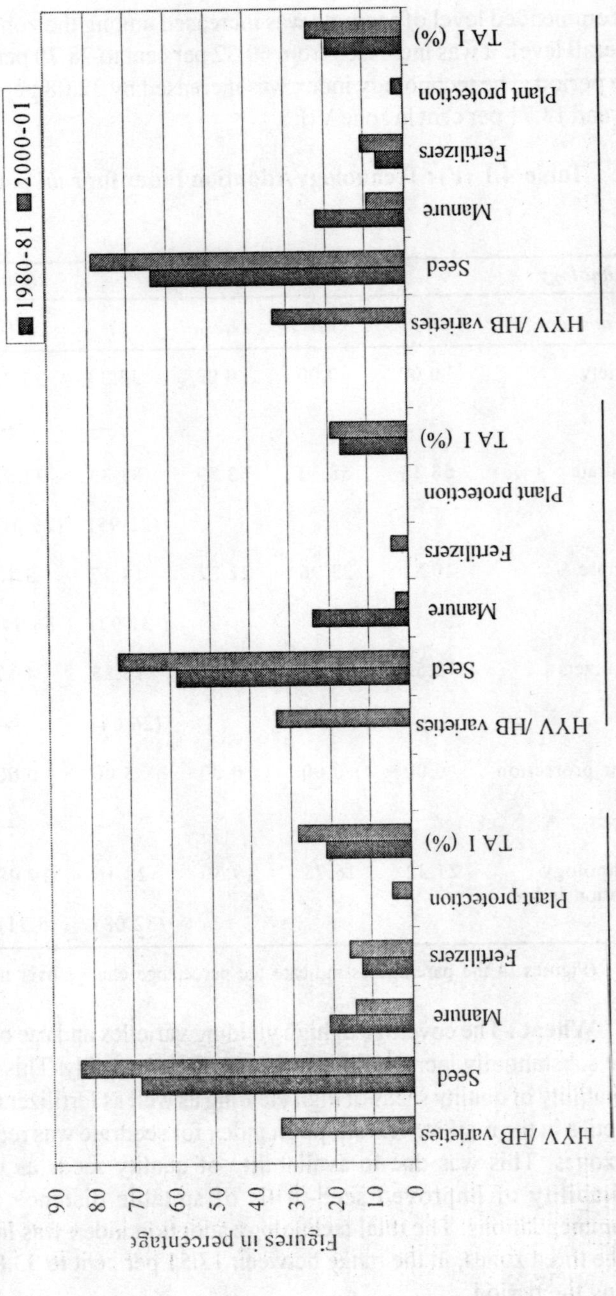

Table 4.1 (G) : Technology Adoption Index for Wheat

(Per cent)

Technology	1980-81				2000-01			
Parameters	*VI*	*VII*	*VIII*	*Overall*	*VI*	*VII*	*VIII*	*Overall*
Variety	5.56	22.22	7.02	11.52	66.67	66.67	66.67	66.67
					(1039.1)	(200.05)	(849.72)	(478.73)
Seedrate	66.67	70.37	63.16	66.67	38.89	37.04	40.35	38.79
					(-41.67)	(-47.36)	(-36.11)	(-41.82)
Manure	12.04	0.00	0.00	3.94	20.37	9.26	8.33	13.03
					(69.19)	—	-	(230.71)
Fertilizers	13.89	8.33	13.16	11.82	27.78	30.56	53.70	37.88
					(100.00)	(266.87)	(308.05)	(220.47)
Plant protection	0.00	11.11	0.00	3.64	5.56	5.56	0.00	3.64
					—	(-49.95)	—	(0.00)
Technology adoption index	19.59	17.84	15.24	17.51	32.16 (64.17)	29.53 (65.53)	36.26 (137.93)	33.01 (88.52)

(Figures in the parentheses indicate the percentage change over the base year)

***Suru* sugarcane :** In case *suru* sugarcane, the total composite index was found to be increased in both the zones and at the overall level during the period under study. This was mainly because of adoption of improved technological components viz; high yielding and fertilizer responsive varieties. The adoption score of seedrate was found to be declined slightly. The improved technology of sugarcane indicated that with improved method of plantations, the seed requirements could be reduced.

The adoption score of manure was found to be increased, excepting in zone VI, wherein, available limited manure was mostly diverted to fruits and vegetables as noticed from other studies.

The *ratoon* sugarcane is supposed to be a residual crop of planted cane. The farmers have no scope to adopt the crucial technology practices viz; variety, spacing, planting method, seedrate, etc. The residual effect of

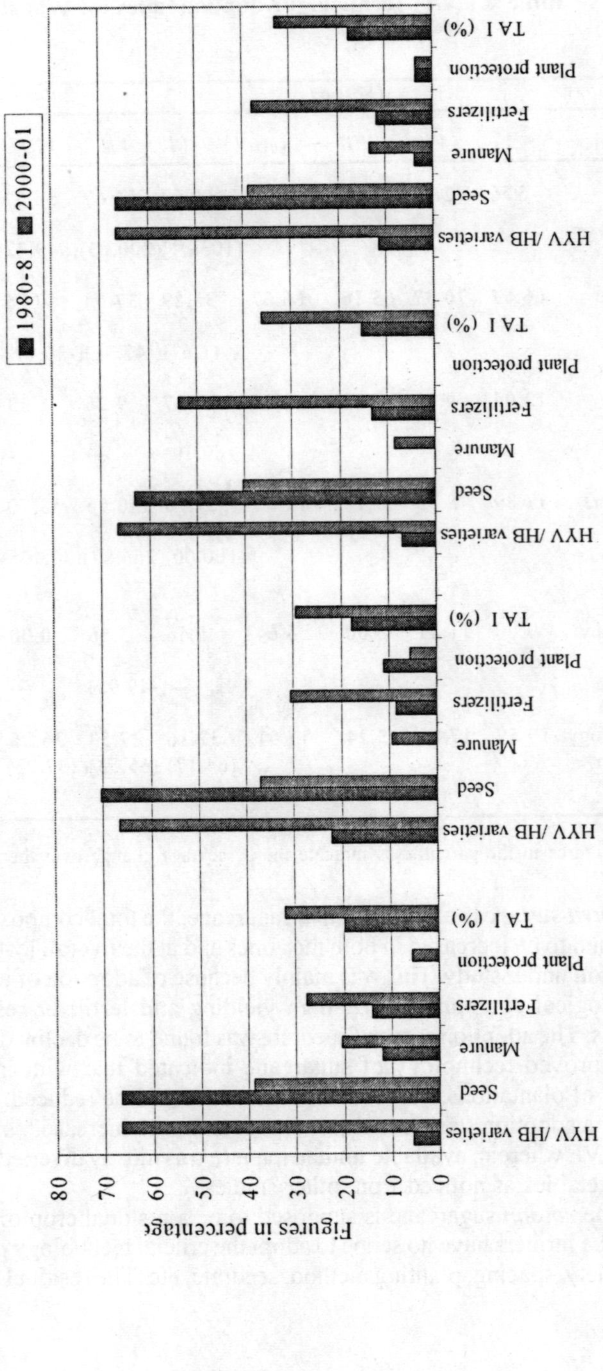

Chart-16: Technology adoption index for wheat in different agro-climatic zones

the manure and fertilizers of the planted cane may also influence the productivity of *ratoon* cane. In view of this, the comparison of the technology adoption index for *ratoon* sugarcane would be misleading. Hence, it is not studied in the present investigation.

Table 4.1 (H) : Technology Adoption Index for *suru* Sugarcane

(Per cent)

Technology	*1980-81*			*2000-01*		
Parameters	*VI*	*VII*	*Overall*	*VI*	*VII*	*Overall*
Variety	34.39	58.12	61.67	66.67	66.67	66.67
				(93.86)	(14.71)	(8.11)
Seedrate	62.60	55.56	58.33	58.33	55.50	57.20
				(-6.82)	(-0.11)	(-1.94)
Manure	34.55	13.25	21.24	19.17	23.61	20.83
				(-44.52)	(78.19)	(-1.93)
Fertilizers	28.46	23.50	25.84	77.50	70.83	75.00
				(172.31)	(201.40)	(190.25)
Plant protection	1.67	0.00	0.01	0.00	0.00	0.00
				—	—	—
Technology adoption index	36.75	33.63	35.68	50.26	49.12	49.84
				(36.76)	(46.06)	(39.69)

(Figures in the parentheses indicate the percentage change over the base year)

The information on technology adoption index revealed that, at the overall level, the technology adoption index ranged in between 9.16 per cent and 35.68 per cent during the year 1980-81. The lowest technology adoption index was observed in pearlmillet. Amongst the five technology parameters, recommended level of seedrate occupied the maximum score. Farmers gave least importance to the plant protection measures, except in the case of cotton. In the year 2000-01, the technology adoption index was in between 24.12 per cent to 49.84 per cent. This indicated that during the period under study, the level of adoption of crop production technology was increased. Similar was the situation in the selected crops across the zones, excepting seedrate manure and plant protection in some cases.

Chart-17: Technology adoption index for sugarcane (suru) in different agro-climatic zones

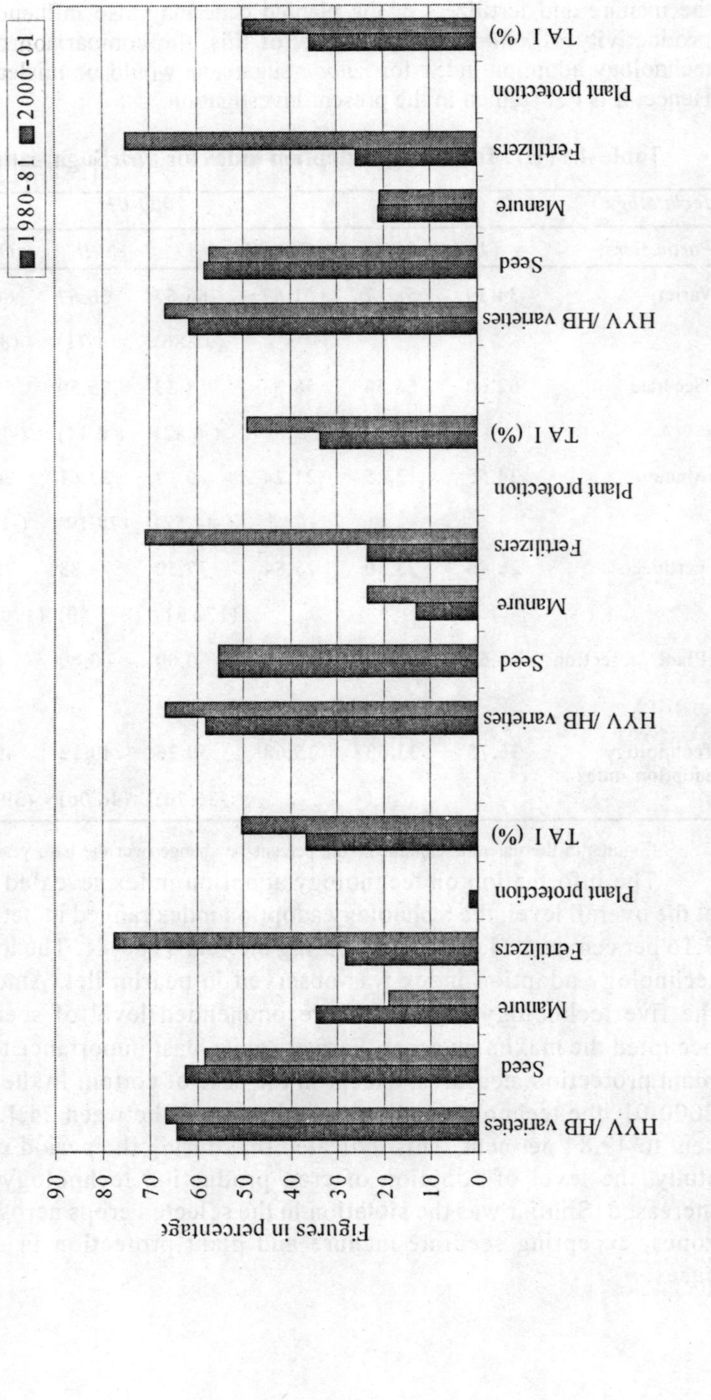

occupied the maximum score. Farmers gave least importance to the plant protection measures, except in the case of cotton. In the year 2000-01, the technology adoption index was in between 24.12 per cent to 49.84 per cent. This indicated that during the period under study, the level of adoption of crop production technology was increased. Similar was the situation in the selected crops across the zones, excepting seedrate manure and plant protection in some cases.

The composite index (C I) of adoption of the technologies based on the adoption index of technology was estimated for each of the farmer separately. The farms were arranged in the ascending order of the composite index of adoption and categorized into three groups based on the range of composite index of adoption viz; low adoption (C·I below 30 %), medium adoption (C I in the range between 31 to 60 %) and high adoption (C I above 61 %). The details of the number of farmers and the percentage change in the Composite Index over the base year are depicted in Appendix- VII. The average composite index of technology adoption at different adoption levels for individual crops and among the agro-climatic zones is indicated in Table 4.2.

At the overall level, the composite indices of technology adoption were at the lower side during the year 1980-81 in all the crops as well as among the agro-climatic zones. During the year 1980-81, the magnitude of composite indices were found in the range between 8.03 to 24.21 per cent, 31.06 to 46.57 per cent and 61.23 to 69.06 per cent under low, medium and high level category of the technology adoption, respectively. While, the figures for the same under respective categories were 17.05 to 28.12 per cent, 37.50 to 55.07 per cent and 73.15 to 78.95 per cent during the year 2000-01. It indicates that, the level of adoption of the technology was increased considerably during last 20 years i.e. 1980-81 to 2000-01.

From the above, it can be summarized that, the technology adoption index was higher in all the agro-climatic zones and for all the crops under study, during the year 2000-01 than that of the year 1980-81.

On an average, the composite index of technology adoption over a period of 20 years was found to be increased to the extent of 29.34 per cent in *rabi* sorghum to 26.76 per cent in *suru* sugarcane. Among the different agro-climatic zones, by and large the composite index of technolog adoption of the components of the improved crop production

Table 4.2 : Average Composite Index (C.I.) of Technology Adoption

(Figures in percentage)

Crop	Zone	Low level adoption		Medium level adoption		High level adoption		Average technology C. I.		
		1980-81	2000-01	1980-81	2000-01	1980-81	2000-01	1980-81	2000-01	% change
Kh. sorghum	VII	18.95	21.05	34.21	39.94	61.93	77.26	27.84	41.59	49.41
	VIII	15.54	26.32	31.57	38.70	—	—	16.30	36.34	112.91
	Overall	16.75	23.69	33.97	39.28	61.93	77.26	22.21	39.03	75.76
Kh. paddy	II	13.74	25.26	31.06	40.63	—	—	14.61	36.79	151.87
	V	18.28	26.29	32.64	41.02	—	—	19.00	35.13	84.90
	IX	18.42	26.32	—	42.11	—	—	18.42	38.95	111.47
	Overall	16.84	25.99	31.85	41.29	—	—	17.34	36.96	113.11
Pearlmillet	V	8.03	19.35	31.58	37.50	—	—	9.21	24.80	169.29
	VI	9.21	20.83	—	39.58	—	—	9.21	24.58	166.88
	VIII	9.02	23.40	—	33.33	—	—	9.02	24.11	167.29
	Overall	8.74	21.13	31.58	37.88	—	—	9.16	24.54	167.88
Groundnut	V	20.00	20.61	31.08	41.23	69.06	78.95	32.06	42.84	33.64
	VII	15.79	17.05	32.56	47.33	61.35	68.42	25.15	42.33	68.33
	Overall	17.11	18.83	31.66	44.62	66.49	76.32	28.60	42.58	48.87

(Contd.)

(Contd.)

Crop	*Zone*	*Low level adoption*		*Medium level adoption*		*High level adoption*		*Average technology C. I.*		
		1980-81	2000-01	1980-81	2000-01	1980-81	2000-01	1980-81	2000-01	% change
Cotton	VII	24.21	28.12	37.59	41.78	—	—	34.07	39.62	16.30
	VIII	19.62	23.99	42.69	45.82	—	73.68	29.55	45.03	52.38
	Overall	21.05	26.47	39.19	43.86	—	73.68	31.75	42.40	32.52
Rabi sorghum	VI	16.78	18.66	34.47	39.77	—	—	20.32	28.16	38.59
	VII	16.78	17.19	33.64	37.37	—	—	17.83	20.97	17.61
	Overall	16.78	17.86	34.30	39.17	—	—	19.21	24.97	29.94
Wheat	VI	16.14	21.64	36.84	42.69	—	—	19.59	32.17	64.19
	VII	17.84	22.53	31.98	43.79	—	—	18.63	31.98	71.69
	VIII	18.24	—	32.83	47.12	—	—	19.01	47.12	147.90
	Overall	17.47	22.11	35.07	45.27	—	—	19.07	37.27	95.41
S.cane (suru)	VI	17.98	21.65	41.11	48.64	66.51	73.15	54.15	68.12	25.80
	VII	23.49	—	46.57	55.07	61.23	69.81	47.61	61.21	28.57
	Overall	20.18	21.65	44.39	53.64	65.27	72.39	51.70	65.53	26.76

technology was found to be increased with the few exceptions. However, the magnitude of the technology adoption index was less than 50 per cent, indicating thereby the constraints in adoption of new technology and suggesting measures for removal of these constraints in all the agro-climatic zones of Maharashtra. This finding was consistent with the results obtained by Pattanaik (1994) who revealed that the large, small and tanent farmers in green belt have adoption index of 87.71, 86.12, and 81.33 per cent, respectively, while, their counterparts in the dryland belt have adoption index of 31.17, 29.67 and 21.54 per cent, respectively. Raut (1997) also confirmed that, overall technology adoption index was increased to 44.88 per cent in irrigated region and 36.17 per cent in rainfed region over the period of 18 years i.e. 1973-74 to 1991-92. Gaddi et al (2002) observed in Dharwad and Bellary districts of Karnataka that, the level of adoption of new technology was found higher during the year 1998-99 than that of in 1996-97 in the case of cotton farms.

In the present investigation, the result indicated that technology adoption index was increased .ı all the crops and among the different agro-climatic zones under study over a period of 20 years i.e. 1980-81 to 2000-01. Thus, hypothesis viz; technology adoption level found increased in all the agro-climatic zones over a period of time, has been confirmed

4.2 Resource Use and Productivity of Major Crops

The nature of Indian agriculture is undergoing a change in the recent years. This has been indicated by number of recent studies. By implication, changing agriculture means, at the micro level a farm is subjected to change in resource use structure over a period of time in response to technological advancement, price change and agricultural infrastructural development. This might have led to change in resource use structure depending upon the extent of technology transformation at the farm level. These changes can be readily visualized by comparing the type of resource-mix employed on the sample farms at two points of time. Accordingly, an attempt has been made to analyse the change in resource use, costs and returns of selected crop enterprises in the different agro-climatic zones over a period. The resource use structure of major selected crops of the sample farms in different agro-climatic zones for two points of time is indicated in Table 4.3 (A) to 4.3 (I). The tabular results of individual crops are graphically shown in Charts- 18 to 26.

***Kharif* sorghum :** The average per hectare resource use levels of human labour, manure and fertilizers were found to be increased by 37.86, 4.60 and 362.41 per cent, respectively at the overall level. As a result, average yield of *kharif* sorghum was found to be increased by 96.29 per cent in 2000-01 over 1980-81. The bullock power and seedrate use levels were decreased over a period of time. The resource use of human labour, bullock power, seed and manure was decreased and the use of only fertilizer was found to be increased in zone VII and average yield was increased by 50 per cent. However, in zone VIII, except bullock power, use of all remaining four resources was found to be increased and as a result, yield level was increased by 138.40 per cent.

***Kharif* paddy :** Paddy crop is cultivated extensively in agro-climatic zones II, V and IX on considerable area. The crop is grown under rainfed condition in high and assured rainfall zones such as II and IX, whereas, in zone V, it is grown under rainfed as well as irrigated conditions to some extent.

From the resource use structure, presented in Table 4.3 (B), it can be seen that at the overall level, the use of human labour and bullock power were declined substantially (53 % and 20 % , respectively) while, the use of manure, fertilizers and seed was increased by 313.46, 163.62 and 104.85 per cent respectively, during the period. The productivity was also doubled during this period.

The use of human labour declined in zones II and V, while it increased by 24.56 per cent in zone IX. The same trend was observed in case of the utilization of bullock power.

The increase in the use of manure and fertilizers was observed in all the zones, however, the percentage increase in application of manure was more in zone II and that for fertilizers in zone IX. The productivity of kharif paddy was increased at higher rate in II zone (115.47 %) followed by V zone (111.39 %) and IX zone (73.93 %). The reduction in use of human labour, bullock power and seedrate were the signs of increasing the level of technology adoption, which attributed to the higher productivity of *kharif* paddy in zone II.

Pearlmillet : Pearlmillet is the second most important cereal crop in the State on the score of its area share. The crop is grown in *kharif* season. The cultivation of pearlmillet is concentrated more in the selected pockets of zones V, VI and VII depending upon soils and climatic factors.

The utilization of human labour, seed, manure and fertilizers were increased by 44.86, 4.80, 319.73 and 1775.00 per cent, respectively, while the use of bullock power declined by 10.49 per cent, at the over level. The proportionate increase in the productivity of pearlmillet was found to the extent of 181.68 per cent during the period.

The use of human labour increased considerably in zones V and VII, while it was slightly increased in zone VI. The utilization of bullock power

Table 4.3 (A) : Per hectare Resource Use and Productivity for kh. sorghum in Different Agro-climatic Zones

Resources	*1980-81*			*2000-01*		
	VII	*VIII*	*Overall*	*VII*	*VIII*	*Overall*
Human labour (mandays)	93.77	50.31	62.26	87.83 (-6.33)	84.10 (67.16)	85.83 (37.86)
Bullock power (pairdays)	18.42	12.33	14.05	14.30 (-22.37)	10.92 (-11.44)	12.39 (-11.81)
Seed (Kgs)	11.47	9.30	9.89	8.37 (-27.03)	10.00 (7.53)	9.29 (-6.07)
Manure (Qtls)	26.97	4.59	10.65	18.17 (-32.63)	5.71 (24.40)	11.14 (4.60)
Fertilizers-NPK (Kgs.)	41.55	2.41	13.01	49.56 (19.28)	68.36 (2736.51)	60.16 (362.41)
Output (Qtl.)	15.30	7.11	9.97	22.95 (50.00)	16.95 (138.40)	19.57 (96.29)

(Figures in parentheses indicate the percentage change over base year)

Chart-18: Resource use levels in kharif sorghum during the year 1980-81 and 2000-01

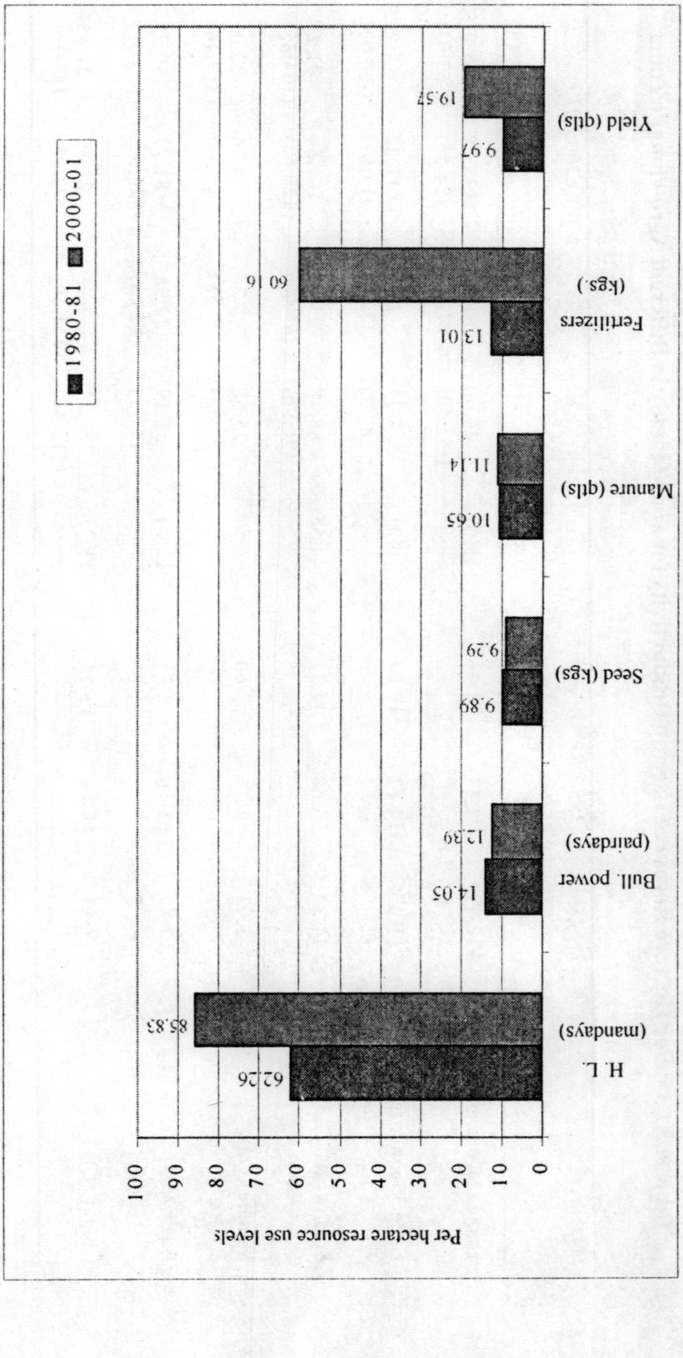

Table 4.3 (B) : Per Hectare Resource Use and Productivity for *kharif* Paddy in Different Agro-climatic Zones

Resources	*1980-81*				*2000-01*			
	II	*V*	*IX*	*Overall*	*II*	*V*	*IX*	*Overall*
Human labour (mandays)	112.49	191.11	111.26	119.03	40.90 (-63.64)	61.82 (-67.65)	138.58 (24.56)	55.91 (-53.03)
Bullock power (pairdays)	11.66	11.81	10.04	10.91	8.31 (-28.73)	7.99 (-32.35)	11.83 (17.83)	8.72 (-20.07)
Seed (Kgs)	44.32	56.68	44.35	44.55	88.18 (98.96)	93.52 (65.00)	101.84 (129.63)	91.26 (104.85)
Manure (Qtls)	3.02	4.15	7.60	5.57	25.68 (750.33)	8.24 (98.55)	8.16 (7.37)	23.03 (313.46)
Fertilizers-NPK (Kgs.)	19.85	68.16	4.25	15.44	22.19 (11.79)	72.95 (7.03)	81.77 (1824.00)	40.75 (163.62)
Output (Qtl.)	9.44	10.80	13.54	11.75	20.34 (115.47)	22.83 (111.39)	23.55 (73.93)	21.35 (81.70)

(Figures in Parentheses Indicate the Percentage Change Over Base Year)

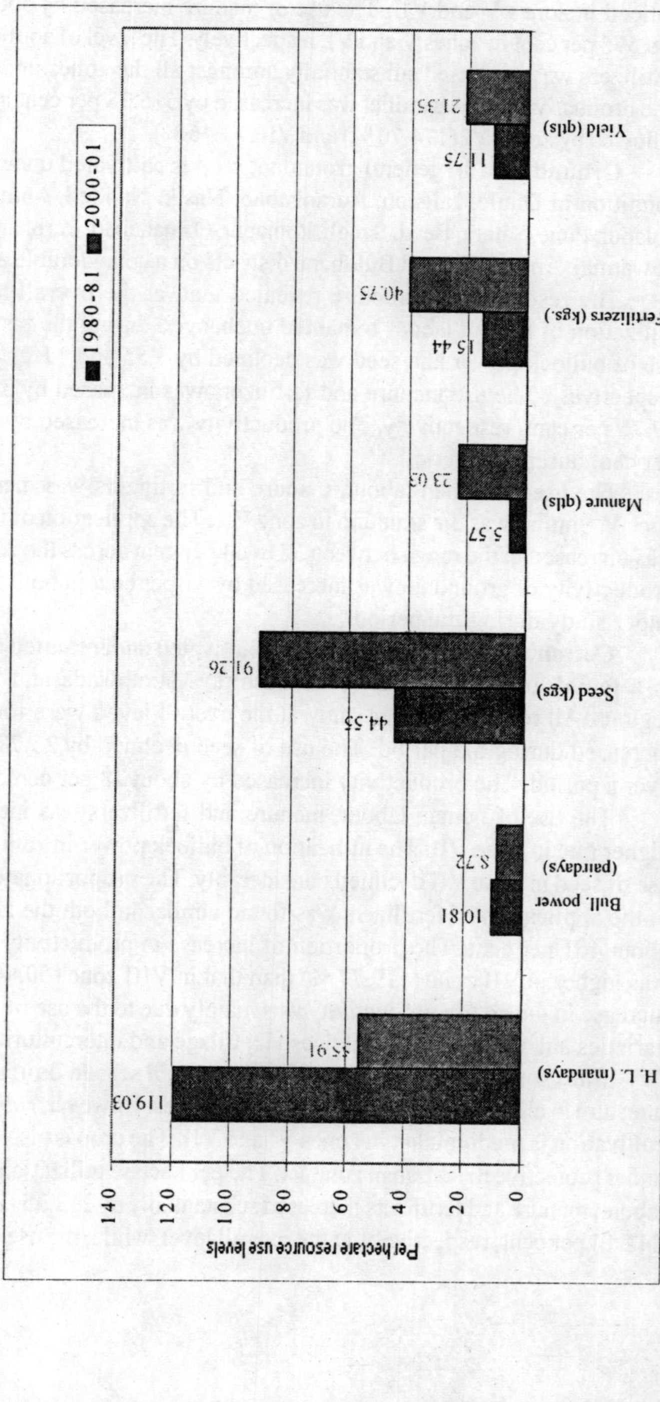

Chart-19: Resource use levels in kharif paddy during the year 1980-81 and 2000-01

was increased by 27.46 per cent in zone V only. The little use of seed was noticed in zones V and VII. The use of manure increased by about 1271 and 575 per cent in zones V and VI, respectively. The level of application of fertilizers was increased substantially amongst all the zones under study. The productivity of pearlmillet was increased by 515.38 per cent in zone V, followed by zones VI (174.70 %) and VII (41.46 %).

Groundnut : In general groundnut crop is cultivated under rainfed condition in Dhule, Jalgaon, Aurangabad, Nasik, Nanded, Ahmednagar, Solapur, Pune, Satara, Beed, Sangli, kolhapur, Osmanabad, Parbhani, Akola, Yewatmal, Amarawati and Buldhana districts on a considerable area.

The resource use structure revealed that, at the overall level, the utilization of human labour remained unchanged during the period. The use of bullock power and seed was declined by 1.55 and 11.21 per cent, respectively. The use manure and fertilizers was increased by 34.67 and 47.75 per cent, respectively. The productivity was increased by about 37 per cent during the period.

The use of human labour, manure and fertilizers was increased in zone V. Similar was the situation in zone VII. The application of fertilizers was increased in the range between 38 to 64 per cent across the zones. The productivity of groundnut was increased by 41 per cent in both the zones under study during the period.

Cotton : Cotton crop is generally cultivated under rainfed condition in a few districts of Western Maharashtra, Marathwada and Vidarbha regions. All resources under study at the overall level, were found to be increased during the period. The use of seed declined by 23.79 per cent over a period. The productivity increased by about 78 per cent.

The use of human labour, manure and fertilizers was increased at higher rate in zone VII. The utilization of bullock power in zone VIII and use of seed in zone VII declined considerably. The proportionate increase in the application of fertilizers was found similar in both the zones, i.e., about 461 per cent. The proportion of increase in productivity of cotton was higher in VII zone (119.77 %) than that in VIII zone (50.96 %). The increase in the cotton productivity was mainly due to the use of improved varieties and management practices viz; tillage and interculturing.

***Rabi* sorghum :** This crop is grown in *rabi* season on residual soil moisture in all the zones, excepting first two zones. However, *rabi* sorghum cultivation is predominant in zones VI and VII. The crop is also cultivated under protective irrigation in zone VI. The per hectare utilization of human labour, manure and fertilizers increased substantially by 213.35, 177.90 and 842.14 per cent, respectively, at the overall level, while, the use of bullock

Table 4.3 (C) : Per Hectare Resource Use and Productivity for Pearlmillet in Different Agro-climatic Zones

	1980-81				2000-01			
	V	VI	VII	Overall	V	VI	VII	Overall
Human labour (mandays)	28.26	43.12	45.97	37.54	69.55 (146.11)	44.94 (4.22)	57.04 (24.08)	54.38 (44.86)
Bullock power (pairdays)	12.60	12.69	12.37	12.77	16.06 (27.46)	10.74 (-15.57)	8.01 (-35.25)	11.43 (-10.49)
Seed (Kgs)	3.57	3.51	3.54	3.54	3.85 (7.84)	3.51 (0.00)	3.97 (12.15)	3.71 (4.80)
Manure (Qtls)	0.35	1.49	4.10	1.47	4.80 (1271.43)	10.05 (574.50)	0.00 —	6.17 (319.73)
Fertilizers-NPK (Kgs.)	1.24	1.89	2.70	1.76	45.58 (3575.81)	22.55 (1093.12)	40.40 (1396.30)	33.00 (1775.00)
Output (Qtl.)	2.60	4.21	6.44	3.93	16.00 (515.38)	9.46 (174.70)	9.11 (41.46)	11.07 (181.68)

(Figures in parentheses indicate the percentage change over base year)

Chart-20: Resource use levels in pearlmillet during the year 1980-81 and 2000-01

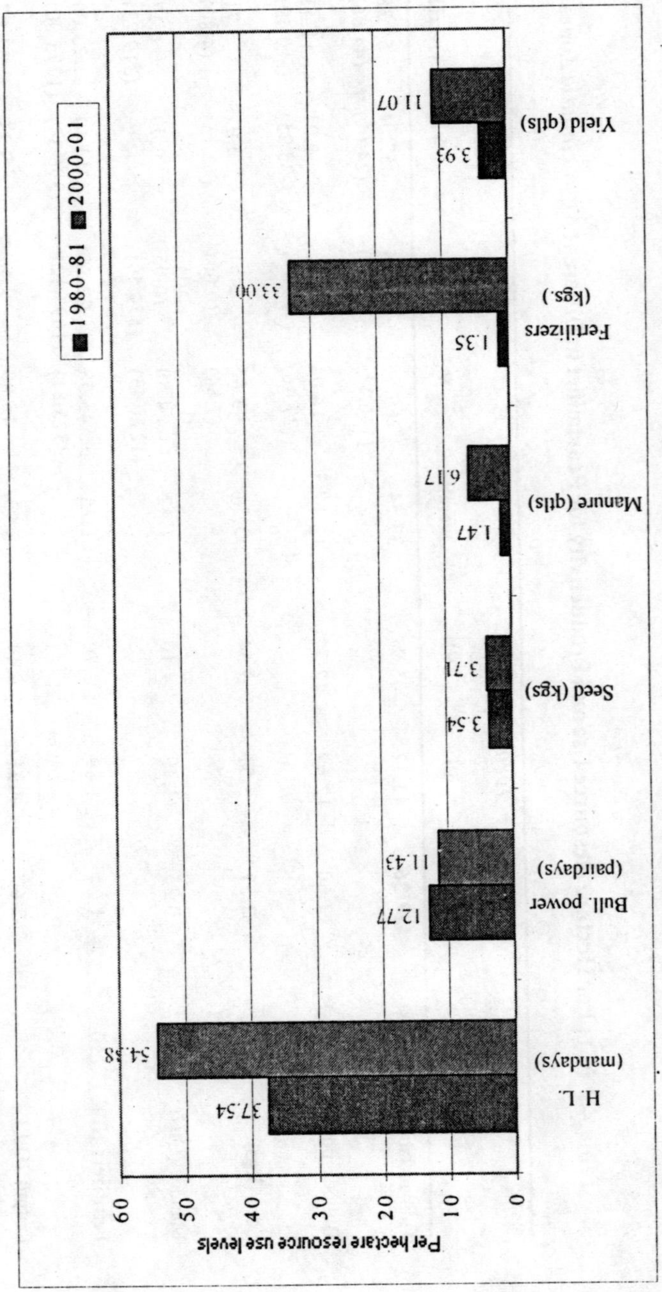

Table 4.3 (D) : Per Hectare Resource Use and Productivity for Groundnut in Different Agro-Climatic Zones

	1980-81			*2000-01*		
	V	*VII*	*Overall*	*V*	*VII*	*Overall*
Human labour (mandays)	100.77	89.11	94.09	121.76 (20.83)	76.48 (-14.17)	94.23 (0.15)
Bullock power (pairdays)	18.48	14.33	16.10	17.47 (-5.47)	14.47 (0.98)	15.85 (-1.55)
Seed (Kgs)	111.80	117.28	114.94	104.45 (-6.57)	100.51 (-14.30)	102.06 (-11.21)
Manure (Qtls)	43.44	14.62	26.62	62.80 (44.57)	18.46 (26.27)	35.85 (34.67)
Fertilizers-NPK (Kgs.)	26.69	34.80	31.33	43.79 (64.07)	47.91 (37.67)	46.29 (47.75)
Output (Qtl.)	13.26	5.66	8.91	18.68 (40.87)	8.00 (41.34)	12.19 (36.81)

(Figures in parentheses indicate the percentage change over base year)

Chart-21: Resource use levels in groundnut during the year 1980-81 and 2000-01

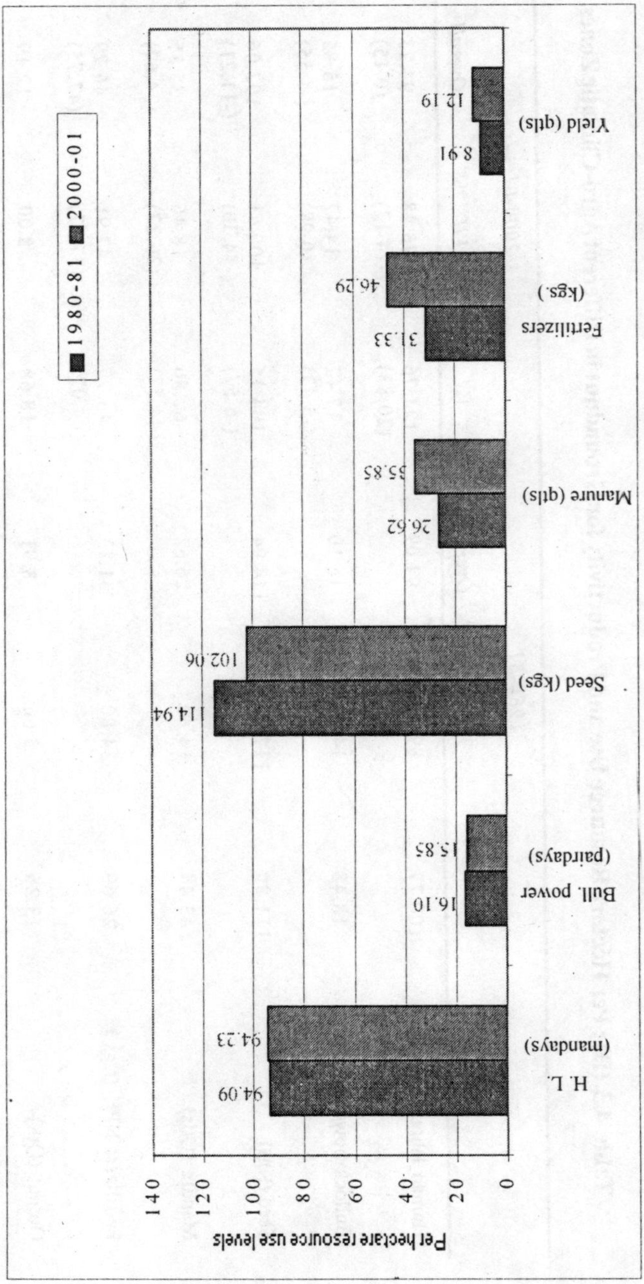

power was declined by about 11 per cent during the period. The little increase in the use of seed quantity was noticed. The productivity was increased by 49.43 per cent.

The per hectare use of human labour was increased by three-times in both the zones during the period, while the application of fertilizers were increased to the tune of 963 per cent in zone VI. The use of manure was increased by 815.84 per cent in zone VI, whereas, the same was declined by 51.30 per cent in zone VII. As a result, the increase in the productivity was found at the higher side in zone VI (by about 92 %) as compared to zone VII (i.e. about 11 %).

Wheat : Wheat crop is cultivated under irrigated condition on large scale in zones VI, VII and VIII. Among the various resources, at the overall level, the utilization of human labour and manure was increased by 29.67 per cent and 143.31 per cent, respectively, while the use of bullock power, seed and fertilizers was declined during the period. The productivity increased was by 31.22 per cent over a period.

The use of human labour increased in all the three zones while, the utilization of bullock power and seed was declined. The use of manure was declined in zone VI, whereas, in zones VII and VIII, the use of this resource was noticed in later period. The application of fertilizers was found to be increased only in zone VIII by about 25 per cent. The increase in the productivity of wheat was observed higher in zone VIII (104.70 %) followed by zone VI (51.99 %) and zone VII (9.03 %).

***Suru* Sugarcane :** The sample from zones VI and VII was studied for sugarcane crop. The sample farmers produced two types of sugarcane viz; plant cane and *ratoon*. In plant cane, *adsali,* pre-seasonal and *suru* plantation is taken up. *Suru* sugarcane is planted on large scale, hence, the analysis for *suru* sugarcane and *ratoon* sugarcane have been attempted separately. The area share of *suru* sugarcane was 23 to 25 per cent and for *ratoon* sugarcane, it was 35 to 40 per cent of the area under total cane during the year 2000-01.

At the overall level, among the resources, the per hectare dose of application of fertilizers was increased by 21.56 per cent only and the productivity was increased by 8.22 per cent during the time.

Amongst the zones, the utilization of human labour, bullock power and seed declined over a period. The use of manure was increased in zone VII only, while, the use of fertilizers was increased in both the two zones under study. The productivity of *suru* sugarcane was increased by only 4.06 per cent in zone VI and 16.62 per cent in zone VII, over a period. Kale et al. (1991) confirmed that there was decrease in the use of human labour,

Table 4.3 (E) : Per Hectare Resource Use and Productivity for Cotton in Different Agro-Climatic Zones

	1980-81			*2000-01*		
	VII	*VIII*	*Overall*	*VII*	*VIII*	*Overall*
Human labour (mandays)	48.29	49.48	48.78	96.69 (100.23)	73.55 (48.65)	82.14 (68.39)
Bullock power (pairdays)	10.96	13.05	11.83	12.53 (14.32)	12.30 (-5.75)	12.39 (4.73)
Seed (Kgs)	5.27	3.00	4.33	3.64 (-30.93)	3.10 (3.33)	3.30 (-23.79)
Manure (Qtls)	3.59	13.10	7.61	19.89 (454.04)	17.83 (36.11)	18.60 (144.42)
Fertilizers-NPK (Kgs.)	15.86	14.75	15.40	87.93 (454.41)	83.67 (467.25)	85.26 (453.64)
Output (Qtl.)	3.44	3.63	3.52	7.56 (119.77)	5.48 (50.96)	6.25 (77.56)

(Figures in parentheses indicate the percentage change over base year)

Chart-22: Resource use levels in cotton during the year 1980-81 and 2000-01

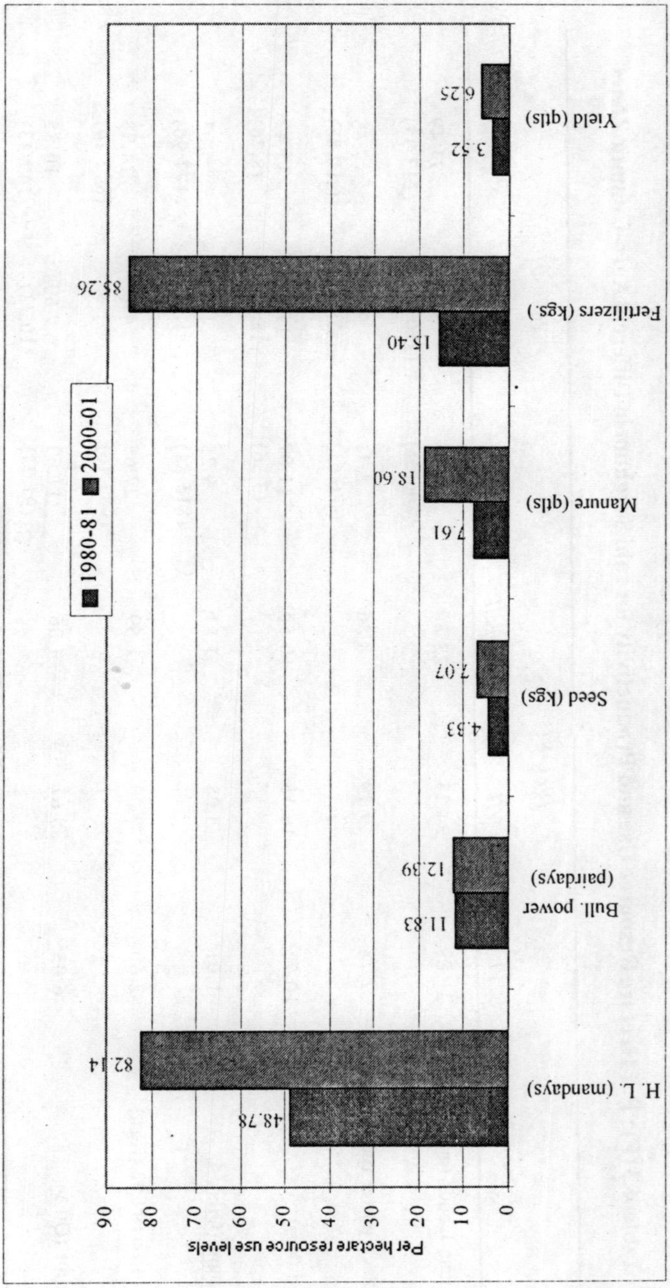

Table 4.3 (F) : Per Hectare Resource Use and Productivity for rabi Sorghum in Different Agro-Climatic Zones

	1980-81			*2000-01*		
	VI	*VII*	*Overall*	*VI*	*VII*	*Overall*
Human labour (mandays)	21.89	24.23	22.80	67.03 (206.21)	75.23 (210.48)	71.49 (213.35)
Bullock power (pairdays)	7.78	9.79	8.56	5.71 (-26.61)	9.24 (-5.62)	7.63 (-10.86)
Seed (Kgs)	10.71	10.18	10.50	11.07 (3.36)	11.71 (15.03)	11.42 (8.76)
Manure (Qtls)	1.01	3.08	1.81	9.25 (815.84)	1.50 (-51.30)	5.03 (177.90)
Fertilizers-NPK (Kgs.)	2.60	0.00	1.59	27.65 (963.46)	4.40 —	14.98 (842.14)
Output (Qtl.)	6.05	8.67	7.06	11.60 (91.74)	9.60 (10.73)	10.55 (49.43)

(Figures in parentheses indicate the percentage change over base year)

Chart-23: Resource use levels in rabi sorghum during the year 1980-81 and 2000-01

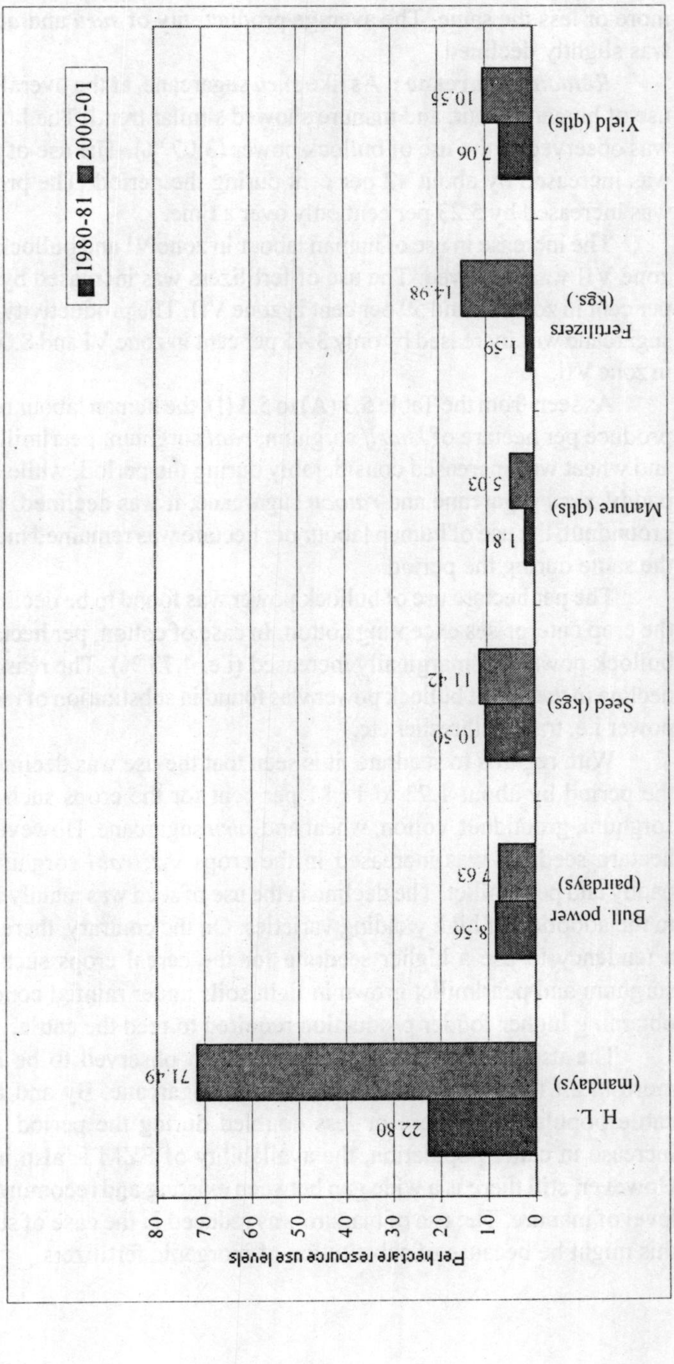

bullock power and manure, whereas, the use of seed material remained more or less the same. The average productivity of *suru* and *adsali* cane was slightly declined.

***Ratoon* Sugarcane :** As like *suru* sugarcane, at the overall level the use of human labour, and manure showed similar trend. The little change was observed in the use of bullock power (3.07 %). The use of fertilizers was increased by about 42 per cent during the period. The productivity was increased by 5.23 per cent only over a time.

The increase in use of human labour in zone VI and bullock power in zone VII was observed. The use of fertilizers was increased by about 47 per cent in zone VI and 59 per cent in zone VII. The productivity of *ratoon* sugarcane was increased by only 3.45 per cent in zone VI and 8.06 per cent in zone VII.

As seen from the Table 5.3 (A) to 5.3 (I), the human labour required to produce per hectare of *kharif* sorghum, *rabi* sorghum, pearlmillet, cotton and wheat was increased considerably during the period, while for *kharif* paddy, *suru* sugarcane and *ratoon* sugarcane, it was declined. In case of groundnut, the use of human labour per hectare was remained more or less the same during the period.

The per hectare use of bullock power was found to be declined for all the crop enterprises excepting cotton. In case of cotton, per hectare use of bullock power has marginally increased (i.e. 4.73 %). The reason for the decline in the use of bullock power was found in substitution of mechanical power i.e. tractor, thresher etc.

With regards to seedrate, it is seen that the use was declined during the period by about 1.72 to 11.81 per cent for the crops such as *kharif* sorghum, groundnut, cotton, wheat and *suru* sugarcane. However, the per hectare seedrate was increased in the crops viz; *rabi* sorghum, *kharif* paddy and pearlmillet. The decline in the use of seed was mainly attributed to the adoption of high yielding varieties. On the contrary, there has been a tendency to use a higher seedrate for the cereal crops such as *rabi* sorghum and pearlmillet grown in light soils under rainfed condition for obtaining higher fodder production required to feed the cattle.

The use of farm-yard manure has been observed to be relatively more in all the crops under study, except sugarcane. By and large, the cattle population is more or less doubled during the period. With the increase in cattle population, the availability of FYM is also increased. However, still there is a wide gap between existing and recommended use level of manure. The use of manure was reduced in the case of sugarcane, this might be because of substitution of inorganic fertilizers.

Table 4.3 (G) : Per Hectare Resource Use and Productivity for Wheat in Different Agro-Climatic Zones

	1980-81				2000-01			
	VI	VII	VIII	Overall	VI	VII	VIII	Overall
Human labour (mandays)	63.08	66.10	63.45	64.35	97.62 (54.76)	91.69 (38.71)	72.40 (14.11)	83.44 (29.67)
Bullock power (pairdays)	14.99	16.21	14.72	15.39	7.69 (-48.70)	5.63 (-65.27)	14.06 (-4.48)	10.28 (-33.20)
Seed (Kgs)	106.44	106.14	109.14	107.04	105.64 (-0.75)	103.62 (-2.37)	105.88 (-2.99)	105.20 (-1.72)
Manure (Qtls)	24.75	0.00	0.00	8.52	21.10 (-14.75)	25.93 —	17.65 —	20.73 (143.31)
Fertilizers-NPK (Kgs.)	114.49	117.69	86.02	108.23	62.38 (-45.51)	77.54 (-34.12)	107.48 (24.95)	89.02 (-17.75)
Output (Qtl.)	15.10	17.83	8.08	14.32	22.95 (51.99)	19.44 (9.03)	16.54 (104.70)	18.79 (31.22)

(Figures in parentheses indicate the percentage change over base year)

Chart-24: Rsource use levels in wheat during the year 1980-81 and 2000-01

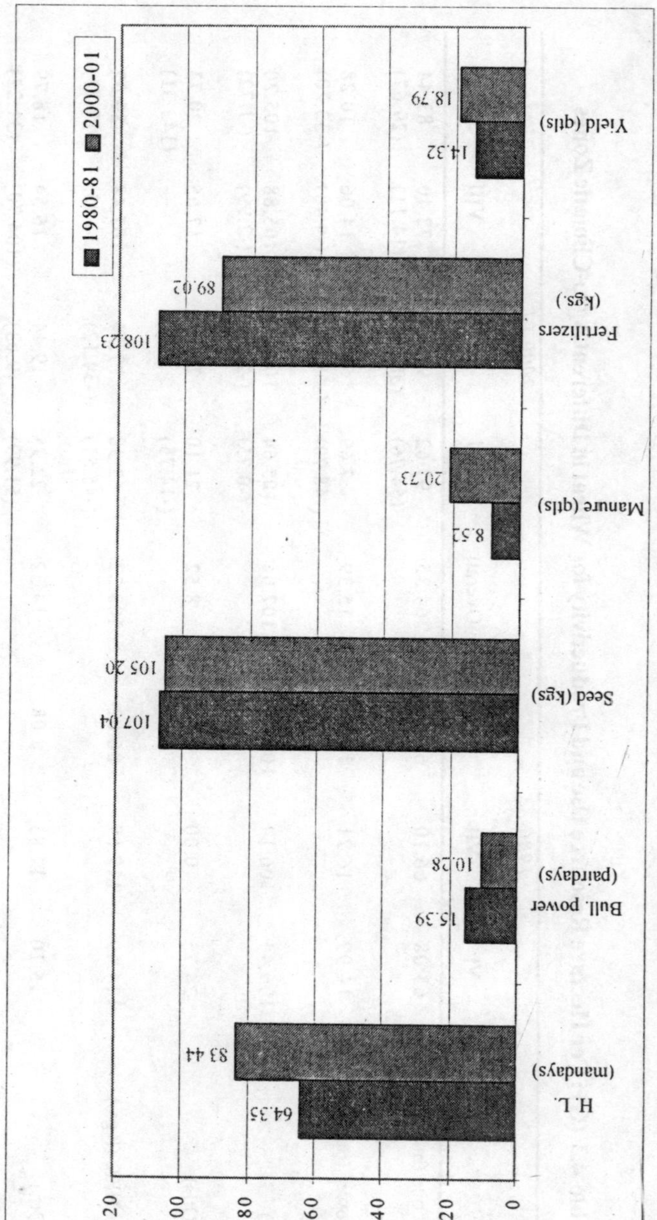

Table 4.3 (H) : Per Hectare Resource Use and Productivity for *suru* Sugarcane in Different Agro-Climatic Zones

	1980-81			*2000-01*		
	VI	*VII*	*Overall*	*VI*	*VII*	*Overall*
Human labour (mandays)	441.02	470.85	450.90	284.95 (-35.39)	277.92 (-40.97)	277.63 (-38.43)
Bullock power (pairdays)	22.62	25.55	23.59	15.27 (-32.49)	13.67 (-46.50)	13.65 (-42.14)
Seed (Kgs)	7759.00	7698.00	7738.38	7015.21 (-9.59)	7462.24 (-3.06)	7214.51 (-6.77)
Manure (Qtls)	114.52	56.50	95.42	41.19 (-64.03)	60.68 (7.40)	52.97 (-44.49)
Fertilizers-NPK (Kgs.)	394.78	329.67	373.20	437.10 (10.72)	466.06 (41.37)	453.65 (21.56)
Output (Qtl.)	943.42	854.75	914.03	981.76 (4.06)	996.78 (16.62)	989.17 (8.22)

(Figures in parentheses indicate the percentage change over base year)

Chart-25: Resource use levels in sugarcane (suru) during the year 1980-81 and 2000-01

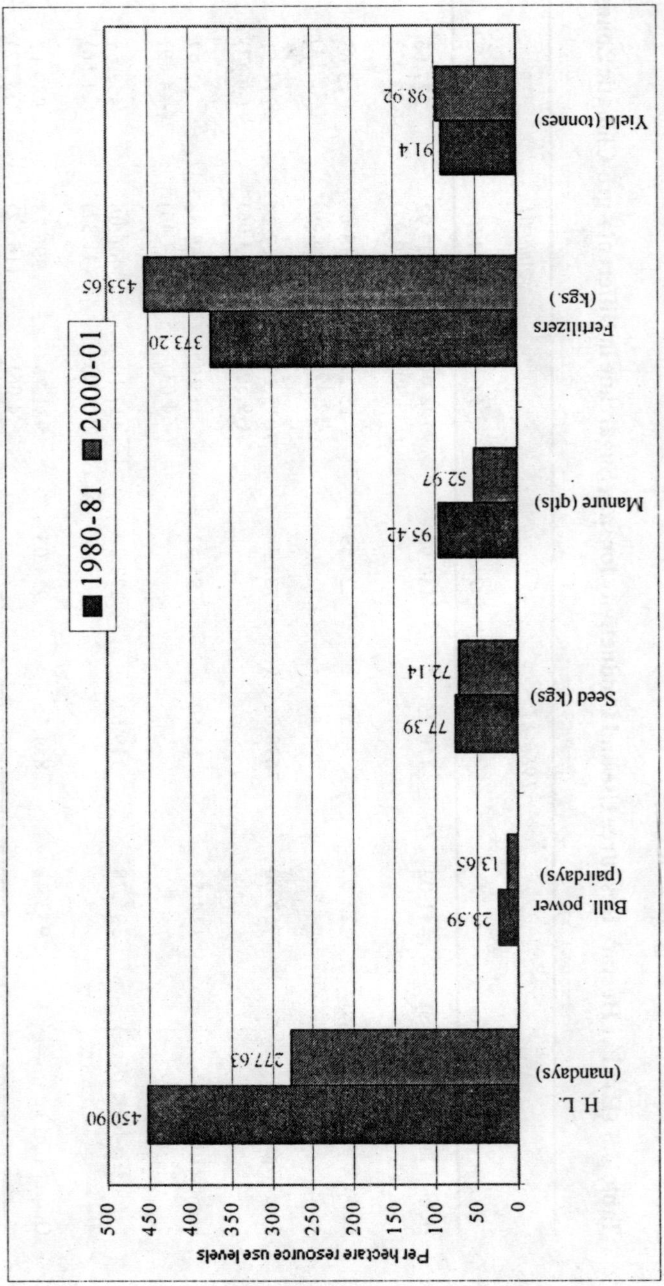

Concerning to the fertilizer resource, it is seen that during the year 1980-81, the use of fertilizers was relatively at lower level, whereas, during the later period the awareness about fertilizer technology was increased, as a result, an increasing tendency to apply fertilizers was observed. Looking to the per hectare quantum of the fertilizers used for different crop enterprises (except sugarcane), it is seen that there existed a great variability in allocating the fertilizer resource among different agro-climatic zones for various crop production enterprises. The allocation of fertilizer resource among different agro-climatic zones for various crop productions seemed to have governed more by productivity and output price expectations, rather than that of any other considerations.

It can be revealed that excepting bullock power and seed resources, the per hectare use of all other resources was relatively higher for the crop enterprises grown in different agro-climatic zones during the later period. The per hectare output of different crops have increased substantially during the period. The rate of increase, however, varied considerably among the crop enterprises. However, in the case of sugarcane, the increase in output was very meagre i.e. only 5.23 per cent to 8.22 per cent.

The findings of the foregoing analysis are consistent with Cavery (1991), Kapase (1992) and Raut (1997). According to Cavery (1991), the productivity of groundnut in Tamil Nadu was increased due to changes in resources use pattern as a consequences of new technology during the period of 22 years (i.e. 1965-66 to 1987-88). Kapase (1992) observed that the per hectare use levels of crucial inputs viz; plant nutrients, human labour, quality seeds, etc. was higher in the zones having assured and high rainfall. The agro-climatic differences among the zones in the State led to wide variations in per hectare resource use structure, costs and returns of individual crops. Raut (1997) also confirmed that the per hectare resource use levels for most of the crops were higher and there was substantial increase in per hectare input utilization during the period from 1973-74 to 1991-92. Thus, the hypothesis viz; because of technological impact, there has been an increase in productivity of the crops under study, over a period of time, has been proved.

4.3 Impact of Technology on Costs and Returns of Major Crops

The structural adjustments in the use of resources bring about changes in the total cost structure to the extent to which these adjustments take place at the farm level. The costs and returns of the individual crop enterprise are also influenced largely depending upon the degree of change taking place in the resource use structure. The quantum and proportion of

Table 4.3 (I) : Per Hectare Resource Use and Productivity for *ratoon* Sugarcane in Different Agro-Climatic Zones

	1980-81			*2000-01*		
	VI	*VII*	*Overall*	*VI*	*VII*	*Overall*
Human labour (mandays)	254.17	308.54	278.48	273.23 (7.50)	225.10 (-27.04)	240.07 (-13.79)
Bullock power (pairdays)	5.26	5.86	5.53	3.90 (-25.86)	7.07 (20.65)	5.70 (3.07)
Seed (Kgs)	—	—	—	—	—	—
Manure (Qtls)	36.58	30.54	33.88	8.80 (-75.94)	27.20 (-10.94)	15.09 (-55.46)
Fertilizers-NPK (Kgs.)	298.72	204.78	256.72	438.98 (46.95)	326.09 (59.24)	364.62 (42.03)
Output (Qtl.)	768.76	718.91	746.47	795.29 (3.45)	781.16 (8.66)	785.49 (5.23)

(Figures in parentheses indicate the percentage change over base year)

Chart-26: Resource use levels in sugarcane (ratoon) during the year 1980-81 and 2000-01

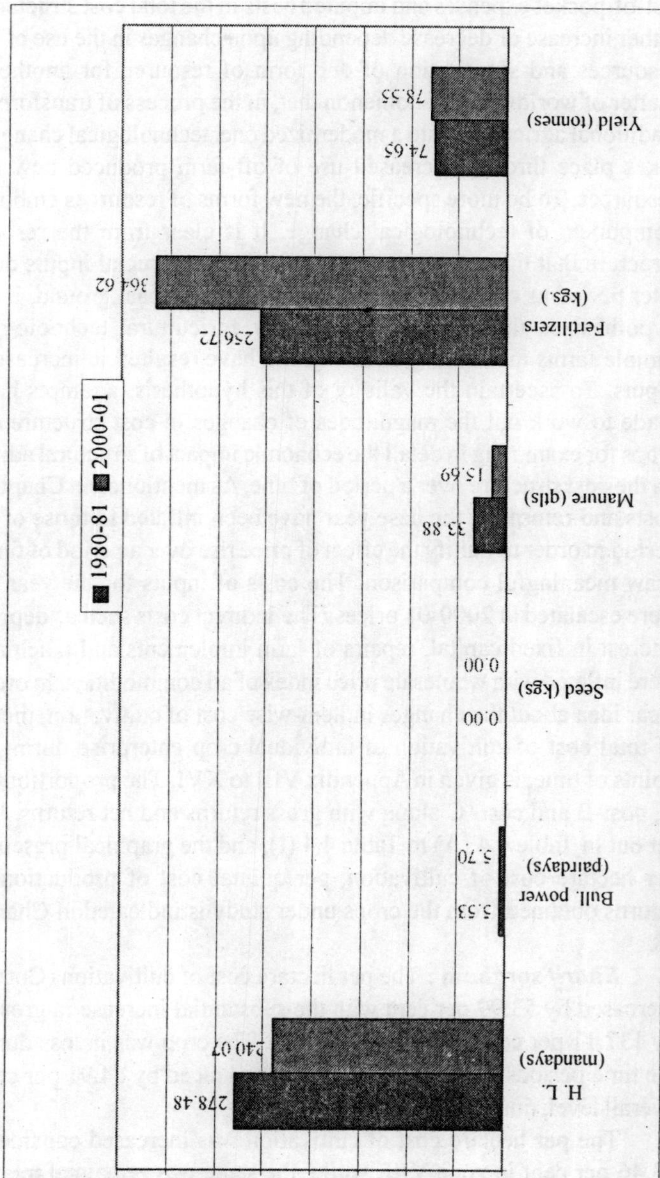

out-of-pocket expenses and imputed costs in the total cost structure which, either increase or decrease depending upon changes in the use of different resources and substitution of one form of resource for another. It is a matter of worldwide phenomenon that, in the process of transformation of traditional agriculture into a modernized one, technological change usually takes place through increased use of off-farm produced new forms of resources. To be more specific, the new forms of resources embody in the component of technological change. It is clear from the resource use structure that there was change in use of major crucial inputs during the later period as compared to base period. On this background, it could be hypothesized that the adoption of new agricultural technology by the sample farms in the study area ought to have resulted in increased use of inputs. To ascertain the validity of this hypothesis, attempts have been made to work out the magnitudes of changes in cost structure of major crops for examining in detail the economic impact of structural adjustments on the cost structure over a period of time. As mentioned in Chapter- 3, the costs and returns of the base year have been inflated to those of the later period in order to nullify the effect of price rise over a period of time and to draw meaningful comparison. The costs of inputs for the year 1980-81 were escalated at 2000-01 prices. The indirect costs such as depreciation, interest in fixed capital, repairs of farm implements and machinery, etc. were inflated with wholesale price index of all commodities. In order to get clear idea about the changes in item-wise cost of cultivation, the breakup of total cost of cultivation of individual crop enterprise during the two points of time, is given in Appendix VIII to XVI. The proportions of cost-A, cost-B and cost- C along with gross returns and net returns, however, set out in Table 4.4 (A) to Table 4.4 (I), and the graphical presentation of per hectare cost of cultivation, per quintal cost of production and net returns obtained from the crops under study is indicated in Charts- 27 to 28.

***Kharif* sorghum :** The per hectare cost of cultivation (Cost- C) was increased by 53.99 per cent with the substantial increase in gross returns by 137.11 per cent, at the overall level. The crop was in loss during both the time periods. However, the loss was reduced by 64.91 per cent at the overall level, during the year 2000-01.

The per hectare cost of cultivation was increased considerably by 93.46 per cent in zone VIII, while, the same was remained more or less similar in zone VII during the period. Similarly, the average per hectare gross returns were higher in zone VII than that in zone VIII. However, change in gross returns were higher in later zone (173.67 %) than the former (67.97 %). The crop was in loss in both the zones, during both the periods, however, the quantum of the loss was reduced by 27.28 per cent in zone VIII and 93.12 per cent in zone VII. It was also observed that the crop was found in loss even at cost-A level in both the zones during the year 1980-81, but the situation was improved in 2000-01, as it was found profitable in both the zones.

***Kharif* paddy :** The average per hectare cost- C at the overall level, was increased marginally (3.03 %), whereas, the gross returns were increased by 69.35 per cent over the period. The paddy cultivation was in loss during 1980-81, while it was found profitable during the year 2000-01. The per hectare net returns were increased from Rs. -1411.89 to Rs. 6282.61 during the period under study. The B C ratio was 0.89 in the base year and the same was worked out to 1.47 in the later year.

Amongst the zones, the per hectare cost- C was increased marginally (3.45 %) in zone II, and it declined in zone V by 15.39 per cent, while the same was increased considerably in zone IX by 74.22 per cent. The gross returns were increased in the range between 74 to 152 per cent among the zones, during the period. The per hectare net returns were substantially increased. The highest increase was in zone II (275.91 %) followed by zone V (200.25 %) and zone IX (74.31 %). The crop was profitable during the later period, with B C ratios in the range of 1.30 to 1.56. The profitability was because of adoption of improved crop production technology by the paddy growers.

Pearlmillet : The per hectare cost of cultivation at the overall level, was increased from Rs. 6008.04 in 1980-81 to Rs 8254.01 in 2000-01. The overall increase in cost- C was to the extent of 37.38 per cent during the period. The gross returns were increased from Rs. 2909.06 in 1980-81 to the tune of Rs. 7695.13 in 2000-01. The crop was in loss during both the time periods, however, the loss was minimised by 81.97 per cent during 2000-01.

The proportion of increase in per hectare cost- C was found the highest in zone V (116.29 %) followed by zone VI (9.82 %) and zone VII (7.93 %). The similar trend was observed in the case of gross returns, as its increase over base period was ranged in 52.53 to 411.15 per cent. The crop was in loss even at level of cost- A, in all the zones under study during 1980-81. The situation was improved during the period in zone V, as the crop was found marginally profitable at cost- C level by Rs. 53.68, while, it was profitable at the level of cost- A in VI and VII zones during the year 2000-01. This was because of better adoption of crop production technology in zone V as compared to other two zones.

Groundnut : At the overall level, the per hectare total cost of cultivation was declined marginally by 3.37 per cent with increase in per hectare gross returns by 36.10 per cent over a period of twenty years. The proportion of per hectare net returns was increased by about 79 per cent. The crop was found in loss during both the periods.

Concerning to the agro-climatic zones, it is seen that the cost of cultivation was remained almost the same in zone V (i.e. about Rs. 26,000), whereas, it was decreased by about 10 per cent in zone VII. The gross returns were increased by 41.92 and 36.54 per cent, respectively, in zones V and VII, during the period. The crop was found in loss during both the periods across the zones, however, the loss was reduced by 92.79 per cent in zone V and 70.41 per cent in zone VII. The B C ratios during 1980-81 and 2000-01 were in the range of 0.57 to 0.73 and 0.86 to 0.98, respectively.

Cotton : The per hectare cost- C and the gross returns at the overall level, were increased by 60.90 and 73.13 per cent, respectively, during the period. The crop was in loss in both the periods, but the loss was minimized to some extent by about 21 per cent during the year 2000-01.

The proportionate increase in the per hectare cost of cultivation as well as gross returns were comparatively higher in zone VII than that in zone VIII. The figures for the same in zone VII were 57.30 and 119.65 per cent, respectively, while, the corresponding figures in the zone VIII were 33.78 and 43.97 per cent over the period. The crop was found profitable only in zone VII, during the year 2000-01, which was because of better output due to more adoption of parameters of improved technology. The B C ratios during the year 1980-81 and 2000-01 were ranged between 0.77 to 0.79 and 0.83 to 1.10, respectively.

***Rabi* sorghum :**At the overall level, the per hectare cost -C was increased by 90.35 per cent with increase in gross returns by 92.87 per cent during the period. The net returns were doubled with the B C ratio at 1.17.

Amongst the zones, the increase in per hectare cost of cultivation and gross returns were considerably higher in zone VI (i.e. 110.18 and 145.71 %) than that in zone VII (69.33 and 44.98 %). This was because of more output due to better adoption of improved technology in zone VI. The substantial increase in the per hectare net returns were noticed in zone VI (1700.08 %), while, the same were declined in zone VII by 28.70 per cent. The crop was found profitable during both the periods, however, the extent of profit over the base year was more in later year.

Wheat : At the overall level, the proportionate decline in the cost of cultivation was observed to the tune of 12.14, whereas, the gross returns were increased by 27.49 per cent. The crop was in loss even in both the

Table 4.4 (A) : Per Hectare Costs and Returns form *kharif* Sorghum in Different Agro-Climatic Zones

(Rupees)

Particulars	*1980-81*			*2000-01*		
	VII	*VIII*	*Overall*	*VII*	*VIII*	*Overall*
Cost- A	10007.62	4795.28	6209.28	9281.6 (-7.25)	8447.15 (76.16)	8811.08 (41.90)
Cost- B	11518.27	5453.95	7099.03	12014.14 (4.30)	10820.23 (98.39)	11340.93 (59.75)
Cost- C	13687.65	6317.69	8316.81	13583.68 (-0.76)	12222.14 (93.46)	12807.23 (53.99)
Gross returns	7847.56	3795.80	4894.99	13181.66 (67.97)	10388.11 (173.67)	11606.47 (137.11)
Net returns	-5840.09	-2521.89	-3421.82	-402.02 (93.12)	-1834.03 (27.28)	-1200.76 (64.91)
B C ratio	0.57	0.60	0.59	0.97	0.85	0.91

(Figures in parentheses indicate the percentage change over base year)

Table 4.4 (B) : Per Hectare Costs and Returns form *kharif* Paddy in Different Agro-Climatic Zones

(Rupees)

	1980-81							
Particulars	*II*	*V*	*IX*	*Overall*	*II*	*V*	*IX*	*Overall*
Cost- A	7113.82	12036.26	6287.85	8498.98	6905.19 (-2.93)	7917.09 (-34.22)	10517.97 (67.27)	7629.09 (-10.24)
Cost- B	8669.11	13590.11	8874.88	10484.85	10226.93 (17.97)	13828.84 (1.76)	17397.71 (96.03)	12027.9 (14.72)
Cost- C	11039.75	18346.25	11252.93	13078.08	11420.62 (3.45)	15522.69 (-15.39)	19605.26 (74.22)	13474.15 (3.03)
Gross returns	7926.77	9637.69	14658.95	11666.19	16896.80 (113.16)	24252.75 (151.64)	25542.19 (74.24)	19756.76 (69.35)
Net returns	-3112.98	-8708.56	3406.02	-1411.89	5476.18 (275.91)	8730.06 (200.25)	5936.93 (74.31)	6282.61 (544.98)
B C ratio	0.72	0.53	1.30	0.89	1.48	1.56	1.30	1.47

(Figures in parentheses indicate the percentage change over base year)

periods, however, loss in the net returns was minimised to the extent of 75.15 per cent during the year 2000-01.

As regards the agro-climatic zones, the decline in per hectare cost-C was to the tune of 12.30 per cent in zone VI followed by zone VIII (5.59 %) and zone VII (0.90 %). On the contrary, the increment in gross returns was higher in zone VIII (97.91 %) followed by zone VI (52.42 %) and zone VII (11.07 %). The crop was found in loss during both the time periods, however, the loss was minimised during later period.

***Suru* sugarcane :** Alike wheat, the per hectare cost of cultivation was declined by 20.10 per cent. The gross returns were increased marginally i.e. 5.45 per cent. However, the per hectare net returns were increased to the tune of 441.16 per cent during the period with the B C ratio to 1.40, at the overall level. The reason for decline in the cost of cultivation could be attributed to the proportionate decline in the use level of crucial inputs such as human labour, bullock power, seed and manure due to the impact of the technology and the marginal increase in the productivity of the crop during the year 2000-01.

The proportion of decline in per hectare cost of cultivation was higher in zone VI (23.82 %) than in zone VII (14.18 %). The gross returns were observed to be increased by 11.94 per cent in zone VII, whereas, the same were marginally increased by about 2 per cent in zone VI, during the period. The crop was found to be profitable in all the zones under study during both the time periods, however, the magnitudes of the B:C ratios indicated that the quantum of profit was more during the year 2000-01. The B:C ratios during the year 1980-81 and 2000-01 were in the range of 1.05 to 1.06 and 1.37 to 1.42.

***Ratoon* sugarcane :** Alike wheat and *suru* sugarcane, at the overall level, the similar trend was noticed in the case of per hectare cost of cultivation. It was declined by 5.24 per cent. The gross returns were increased marginally by 3.48 per cent during the period. The net returns were increased by about 19 per cent, with the B C ratio to 1.71.

The proportion of per hectare cost of cultivation was reduced marginally by 5 to 6 per cent in both the zones, whereas, marginal increment in the gross returns were found in both the zones under study and an increase was 3.13 and 5.01 per cent in the zones VI and VII, respectively. The proportion of an increase in per hectare net returns was 20 to 23 per cent in both the zones. The magnitudes of the B:C ratios were quite at higher side during the year 2000-01.

The data presented in Table 4.4 (A) to 4.4 (I) revealed that, in the case of *kharif* sorghum, *kharif* paddy, pearlmillet, cotton and *rabi* sorghum,

Table 4.4 (C) : Per Hectare Costs and Returns from Pearlmillet in Different Agro-Climatic Zones

(Rupees)

Particulars	1980-81				2000-01			
	V	VI	VII	Overall	V	VI	VII	Overall
Cost- A	4078.04	4602.99	4942.55	4446.50	7100.9 (74.12)	4699.55 (2.10)	5062.32 (2.42)	5413.58 (21.75)
Cost- B	4535.82	5718.73	5769.70	5033.94	9887.01 (117.98)	5983.67 (4.63)	6669.28 (15.59)	7167.88 (42.39)
Cost- C	5414.21	6207.21	6937.01	6008.04	11710.55 (116.29)	6816.88 (9.82)	7486.93 (7.93)	8254.01 (37.38)
Gross returns	2301.56	3027.59	4066.69	2909.06	11764.23 (411.15)	6305.05 (108.25)	6202.76 (52.53)	7695.13 (164.52)
Net returns	-3112.65	-3179.62	-2870.32	-3098.98	53.68 (101.72)	-511.83 (83.90)	-1284.17 (55.26)	-558.88 (81.97)
B C ratio	0.43	0.49	0.59	0.48	1.00	0.92	0.83	0.93

Figures in parentheses indicate the percentage change over base year)

Table 4.4 (D) : Per Hectare Costs and Returns from Groundnut in Different Agro-Climatic Zones

(Rupees)

Particulars	*1980-81* *V*	*VII*	*Overall*	*2000-01* *V*	*VII*	*Overall*
Cost- A	19617.54	12792.68	15709.73	18369.88 (-6.36)	10670.3 (-16.59)	13689.65 (-12.86)
Cost- B	23455.90	14764.69	18479.45	23983.2 (2.25)	13216.91 (-10.48)	17438.86 (-5.63)
Cost- C	25153.39	16127.23	19985.24	26653.09 (5.96)	14574.74 (-9.63)	19311.20 (-3.37)
Gross returns	18439.90	9165.24	13129.37	26169.16 (41.92)	12514.62 (36.54)	17869.18 (36.10)
Net returns	-6713.49	-6961.99	-6855.87	-483.93 (92.79)	-2060.12 (70.41)	-1442.02 (78.97)
B C ratio	0.73	0.57	0.66	0.98	0.86	0.93

(Figures in parentheses indicate the percentage change over base year)

the per hectare cost of cultivation was increased during the period and this increase was in the range of 3.03 to 90.35 per cent. The lowest and the highest being, respectively, for *kharif* paddy and *rabi* sorghum. The per hectare cost of cultivation, in fact, was declined in the case of groundnut, wheat, *suru* sugarcane and *ratoon* sugarcane. This decline was in the range of 3.57 to 20.10 per cent. Based on the relative changes in per hectare cost of cultivation of different crop enterprises it can be said that, by adoption of improved crop production technology, the farmers in different agro-climatic zones could increase the resource use levels and thereby crop productivity and the gross returns of almost all the crops. However, that for sugarcane, it was remained more or less unchanged.

According to Raut (1997) the cost of crop production along with productivity for all crops under study in irrigated and rainfed region was increased during the period of 17 years i.e. 1973-74 to 1991-92 due to the effect of new crop production technology. Rahman and Rahman (1999) found that technological change played a significant role in increase in the cost of crop production as well as crop productivity by examining the regionwise databases of two time series i.e. for 47 years (1948-94) and for 29 years (1961-92). Nagure (2002) found that per hectare cost of cultivation of *rabi* sorghum, wheat and cotton was increased by 102.82, 55.73 and 75.53 per cent, respectively, during the period of 16 years. The findings of this study are consistent with the results of Raut (1997), Rahman and Rahman (1999) and Nagure (2002). Thus, the hypotheses viz; there existed positive change in the costs and returns of the crops under study over a period, has been proved.

4.4 Alternative Measures of Efficiency of Crop Production

It is clear from the earlier discussion that due to adoption of improved crop production technology, there has been simultaneous increase in gross returns alongwith increased costs of cultivation resulting from the structural adjustments in the use of farm resources during the period. The very fact, the gross returns increased at a faster rate than cost of cultivation, which indicated that, the structural adjustment in the use of farm resources have contributed for increased profitability of crop production business. These changes need further examination to understand improvements in the efficiency in crop production business. So far, the measures of farm-efficiency such as per hectare gross returns, net returns etc. have sufficiently supported the fact that the farm efficiency has improved during the period.

An attempt was made to estimate other alternative measures of farm efficiency such as gross returns per manday, paid out cost per unit of

Table 4.4 (E) : Per Hectare Costs and Returns from Cotton in Different Agro-Climatic Zones

(Rupees)

Particulars	*1980-81*			*2000-01*		
	VII	*VIII*	*Overall*	*VII*	*VIII*	*Overall*
Cost- A	7223.47	7913.74	7509.42	10041.57 (39.01)	9813.96 (24.01)	9898.51 (31.81)
Cost- B	8536.92	9201.71	7826.09	13352.55 (56.41)	12521.53 (36.08)	12830.21 (63.94)
Cost- C	9381.00	10170.59	8721.87	14756.77 (57.30)	13606.68 (33.78)	14033.88 (60.90)
Gross returns	7419.36	7833.29	7590.84	16296.86 (119.65)	11277.44 (43.97)	13141.93 (73.13)
Net returns	-1961.64	-2337.30	-1131.03	1540.09 (178.51)	-2329.24 (0.34)	-891.95 (21.14)
B C ratio	0.79	0.77	0.87	1.10	0.83	0.94

(Figures in parentheses indicate the percentage change over base year)

Table 4.4 (F) : Per Hectare Costs and Returns from *rabi* Sorghum in Different Agro-Climatic Zones

(Rupees)

	1980-81			*2000-01*		
Particulars	*VI*	*VII*	*Overall*	*VI*	*VII*	*Overall*
Cost- A	3408.98	3341.98	3383.05	5305.77 (55.64)	5155.91 (54.28)	5223.88 (54.41)
Cost- B	4277.36	4602.15	4403.07	7834.52 (83.16)	7502.9 (63.03)	7653.92 (73.83)
Cost- C	4453.82	5032.97	4677.98	9361.18 (110.18)	8522.57 (69.33)	8904.48 (90.35)
Gross returns	4555.61	6696.87	5384.37	11193.43 (145.71)	9708.87 (44.98)	10384.95 (92.87)
Net returns	101.79	1663.90	706.39	1832.25 (1700.08)	1186.30 (-28.70)	1480.47 (109.58)
B C ratio	1.02	1.33	1.15	1.20	1.14	1.17

(Figures in parentheses indicate the percentage change over base year)

Table 4.4 (G) : Per Hectare Costs and Returns from Wheat in Different Agro-Climatic Zones

(Rupees)

Particulars	*1980-81*				*2000-01*			
	VI	*VII*	*VIII*	*Overall*	*VI*	*VII*	*VIII*	*Overall*
Cost- A	15586.35	12988.46	10542.74	13237.43	10387.95	10523.53	9503.69	9986.11
					(-33.35)	(-18.98)	(-9.86)	(-24.56)
Cost- B	18205.05	16660.88	11915.05	15548.10	14774.31	14712.88	11431.39	13095.54
					(-18.84)	(-11.69)	(-4.06)	(-15.77)
Cost- C	19372.25	16995.38	13506.35	16892.84	16989.55	16842.12	12750.79	14841.56
					(-12.30)	(-0.90)	(-5.59)	(-12.14)
Gross returns	10865.20	13212.90	5504.80	10369.29	16560.69	14676.29	10894.77	13220.24

Chart-27: Changes in cost of cultivation of major crops during 1980-81 and 2000-01

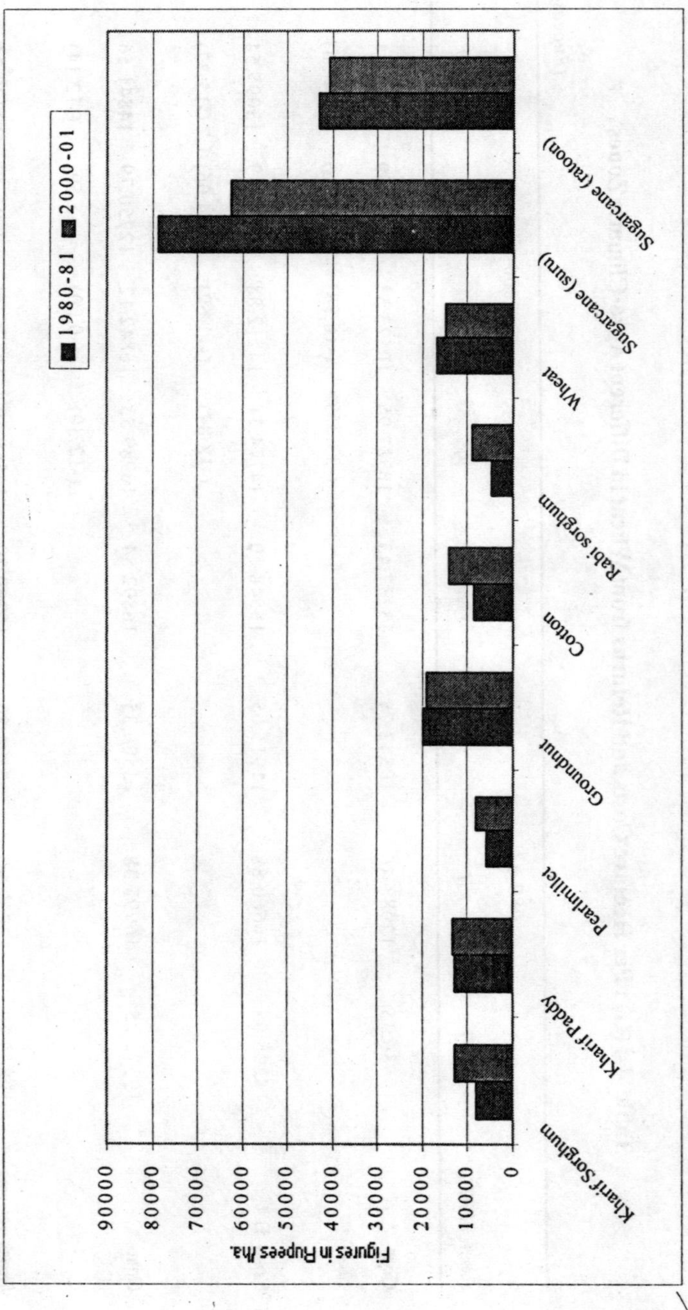

Chart-28: Changes in per quintal cost of production of major crops during the year 1980-81 and 2000-01

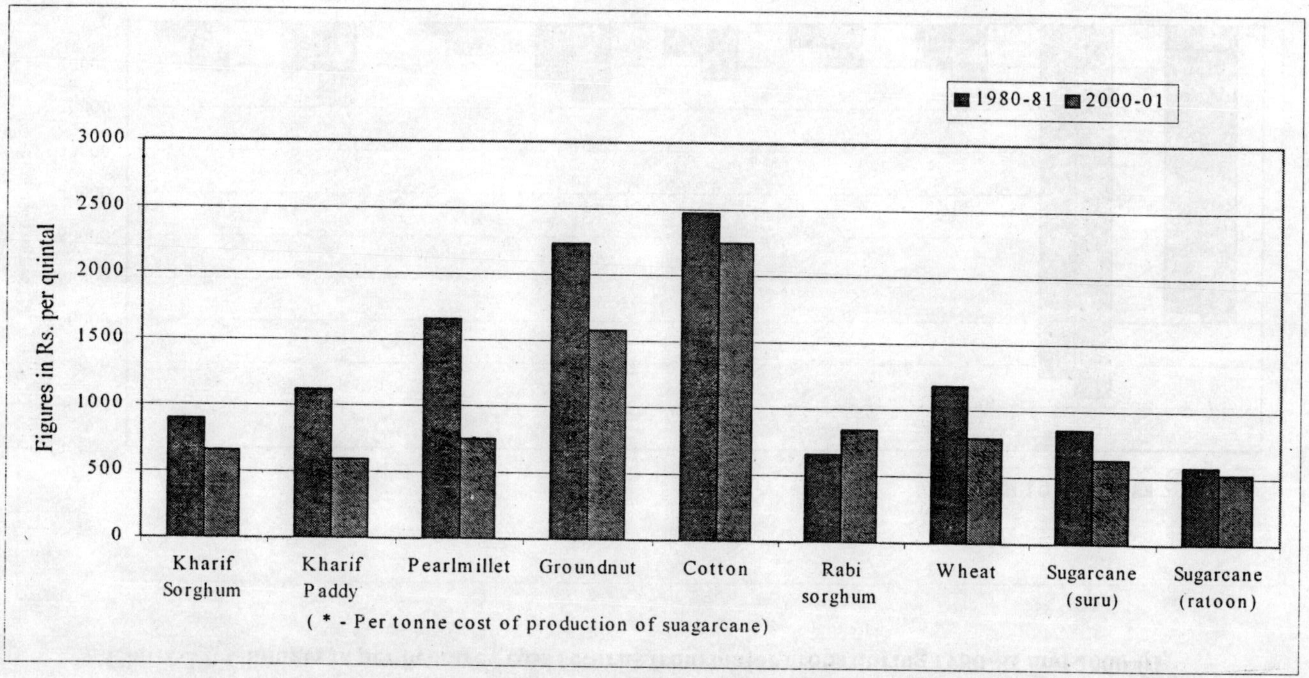

Chart-29: Changes in per hectare gross returns from major crops during 1980-81 and 2000-01

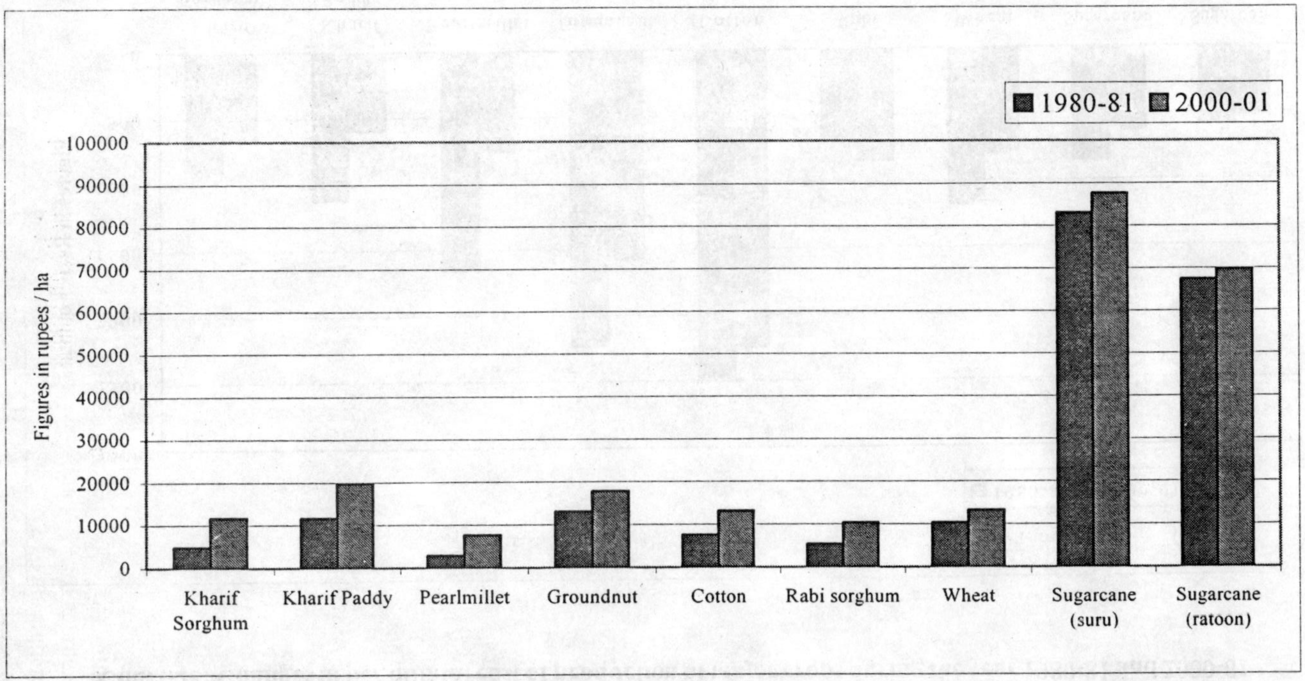

output, total cost per unit of output and gross output per unit of total cost for the years 1980-81 and 2000-01 for all the crops under different agro-climatic zones. The percentage change in the alternative measures of efficiency of crop production on sample farms during 1980-81 and 2000-01 is also shown in the Table 4.5.

From the estimates of alternative measures of efficiency, it can be revealed that there has been significant improvement in the efficiency in managing the crop production business. The percentage change in the alternative measures of efficiency of crop production on sample farms indicated that, at the overall level, the value of gross output per manday was increased in *kharif* sorghum, *kharif* paddy, pearlmillet, groundnut, *suru* sugarcane and *ratoon* sugarcane by 106.96, 260.54, 82.61, 35.90, 71.31 and 20.04 per cent, respectively, during the period. The same was remained more or less unchanged in case of wheat and cotton. The paid-out cost per unit of output was reduced in all the crops under study, except in sugarcane crop. This indicated the lowering down of the farm- labour efficiency. The total cost per unit of output was declined in all the crops, except *rabi* sorghum. This decline was in the range of 9.40 to 54.95 per cent. In *rabi* sorghum, the total cost per unit of output was increased by 27.38 per cent.

The percentage change in gross output per manday was found to be positive in almost all crops among the different agro-climatic zones. However, paid out cost per unit of output was found to be negative in almost all crops gown in different agro-climatic zones excepting both the types of sugarcane. Similar trend was observed in case of per unit gross output. There was positive change in gross output per unit of total cost.

The reduction in per unit of output cost during the period, could be attributed to the increase in the productivity of the crops due to impact of new technology. Similarly, the various ratios in respect of costs and output indicated definite sign of improvement in the farm efficiency during the period.

4.5 Assessment of Returns Due to New Technology

The basic results obtained with the tabular analysis, showed the considerable increment in the outputs of the crops during the period. This increment was due to increase in the adoption level of new crop production technology, which was obviously in conformity with the hypothesis that technical change found to be a major source of output growth. However, as argued by Bisaliah (1997), technical change is not a 'manna from heaven.' Resources must be devoted for that and one should know the costs and returns to producing new technology. Basically, there are three quantitative methods for evaluation of returns at the cost of new technology, (i) Value of input saved approach, (ii) Consumer surplus approach and (iii) Marginal product approach.

Table 4.4 (H) : Per Hectare Costs and Returns from *suru* Sugarcane in Different Agro-Climatic Zones

(Rupees)

Particulars	*1980-81*			*2000-01*		
	VI	*VII*	*Overall*	*VI*	*VII*	*Overall*
Cost- A	57098.47	56128.35	56776.97	40324.56 (-29.38)	42645.52 (-24.02)	41469.38 (-26.96)
Cost- B	71994.51	70065.15	71355.11	56603.66 (-21.38)	59825.39 (-14.61)	58192.79 (-18.45)
Cost- C	80224.04	74981.04	78486.47	61116.84 (-23.82)	64346.92 (-14.18)	62710.09 (-20.10)
Gross returns	85184.63	78768.62	83088.33	87068.22 (2.21)	88173.53 (11.94)	87613.42 (5.45)
Net returns	4960.59	3787.58	4601.86	25951.38 (423.15)	23826.61 (529.07)	24903.33 (441.16)
B C ratio	1.06	1.05	1.06	1.42	1.37	1.40

(Figures in parentheses indicate the percentage change over base year)

Table 4.4 (I) : Per Hectare Costs and Returns from *ratoon* Sugarcane in Different Agro-Climatic Zones

(Rupees)

Particulars	*1980-81* VI	VII	*Overall*	*2000-01* VI	VII	*Overall*
Cost- A	26893.64	26598.54	26761.88	23199.15 (-13.74)	20876.22 (-21.51)	21587.5 -19.33
Cost- B	39090.94	38427.54	38794.36	36185.35 (-7.43)	35886.11 (-6.61)	35977.74 -7.26
Cost- C	43166.45	43042.79	43111.17	40177.78 (-6.92)	41151.76 (-4.39)	40853.18 -5.24
Gross returns	68569.57	66105.97	67468.18	70713.69 (3.13)	69420.08 (5.01)	69816.18 3.48
Net returns	25403.12	23063.18	24357.01	30535.91 (20.21)	28268.32 (22.57)	28963.00 18.91
B C ratio	1.59	1.54	1.56	1.76	1.69	1.71

(Figures in parentheses indicate the percentage change over base year)

Given nature of data, the value of input saved approach is used in the present study to estimate the resources required to produce the per hectare new technology level of output by old technology. The difference between this figure and the resources actually used to produce the new technology level of output represents the value of input saved because of the higher level of efficiency due to new technology. To estimate the value of resources saved with the introduction of new technology and the quantity of extra output produced during the period, the price effect was taken in to account. The following expressions have been formulated to estimate the value of resources saved with the introduction of new technology.

Yo = Per hectare output of crop in quintal during the year 1980-81

Yn = Per hectare output of crop in quintal during the year 2000-01

Ro = Value of labour, fertilizers and capital inputs at constant prices required to produce one unit of Yo with old technology.

Rn = Value of labour, fertilizers and capital inputs at constant prices used in producing one unit of Yn.

r = Rate of change in cost of labour, fertilizers and capital inputs per unit of output over the base period.

Sr = Value of labour, fertilizers and capital inputs saved per unit of output with new technology.

Therefore, Sr = (r/100) Rn

The results on the value of inputs saved in cultivating the crops in different agro-climatic zones are shown in Table 4.6.

At the overall level, the per quintal value of input saved due to new crop production technology was observed to be the highest in groundnut (Rs. 4.65), while, the same was the lowest in *ratoon* sugarcane (Re. 0.05).

In the case *rabi* sorghum, it was noticed that there was no saving in the cost of inputs due to adoption of new technology, as the magnitude of 'Sr' turned out negative.

Across the agro-climatic zones, the saving in the cost of labour and capital inputs due to the new technology was observed in all the crops, excepting in *kharif* paddy in zone IX and *rabi* sorghum in zones VI and VII. In sugarcane crop, saving in cost of inputs was marginal. This indicated that, the new technologies did not influence much the output of this crop. The magnitude of resource saving was due to an upward shift in the production function or a downward shifts on the unit cost of function with the introduction of new technology.

Table 4.5 : Alternative Measures of Efficiency of Crop Production on Sample Farms in 1980-81 and 2000-01

(Figures in rupees)

	Zone	Gross returns per manday			Paid out cost per unit of output			Total cost per unit of output			Gross output per unit of total cost		
		1980-81	2000-01	% change	1980-81	2000-01	% change	1980-81	2000-01	% change	1980-81	2000-01	*% change*
Kh. sorghum	VII	83.69	150.08	79.33	640.62	93.84	-85.35	894.62	591.01	-33.94	0.57	0.97	70.25
	VIII	75.75	123.52	63.06	717.75	87.59	-87.80	888.56	721.07	-18.85	0.60	0.85	41.66
	Overall	65.34	135.23	106.96	687.36	90.28	-86.87	891.45	654.04	-26.63	0.52	0.90	73.47
Kh. paddy	II	70.47	413.12	486.24	863.32	150.27	-82.59	1113.05	561.49	-49.55	0.73	1.48	102.67
	V	50.43	392.31	677.93	824.96	104.33	-87.35	1698.73	520.37	-69.37	0.53	1.56	194.79
	IX	131.75	184.31	39.90	518.06	61.73	-88.09	831.09	832.50	0.17	1.32	1.30	-1.30
	Overall	98.01	353.37	260.54	727.33	117.04	-83.91	1113.36	587.11	-47.27	1.00	1.47	46.63
Pearlmillet	V	81.44	169.15	107.70	1487.06	84.23	-94.34	2082.39	731.91	-64.85	0.48	1.00	109.29
	VI	70.21	140.30	99.83	1046.40	92.98	-91.11	1474.35	720.60	-51.12	0.55	0.92	68.17
	VII	88.46	108.74	22.93	716.92	73.78	-89.71	1077.19	821.84	-23.71	0.66	0.83	25.53
	Overall	77.49	141.51	82.61	1056.26	84.99	-91.95	1655.11	745.62	-54.95	0.55	0.93	69.51
Groundnut	V	182.99	214.92	17.45	1152.61	132.49	-88.51	1896.55	1426.82	-24.77	0.84	0.98	16.89
	VII	102.85	163.63	59.10	1878.42	123.57	-93.42	2849.33	1821.84	-36.06	0.65	0.86	32.10
	Overall	139.54	189.63	35.90	1412.28	128.10	-90.93	2243.01	1584.18	-29.37	0.76	0.93	21.75
Cotton	VII	153.64	168.55	9.70	1645.60	89.66	-94.55	2727.03	1951.95	-28.42	0.95	1.10	16.25
	VIII	150.98	153.33	1.56	1618.71	114.27	-92.94	2801.81	2482.97	-11.38	0.80	0.83	3.60

Crop	Zone	Gross returns per manday			Paid out cost per unit of output			Total cost per unit of output			Gross output per unit of total cost		
		1980-81	2000-01	% change	1980-81	2000-01	% change	1980-81	2000-01	% change	1980-81	2000-01	% change
Rabi sorghum	VI	208.11	166.99	-19.76	414.35	67.72	-83.66	736.17	807.00	9.62	1.16	1.20	3.08
	VII	276.39	129.06	-53.31	315.52	57.68	-81.72	580.50	887.77	52.93	1.47	1.14	-22.50
	Overall	236.16	145.26	-38.49	360.61	61.97	-82.81	662.60	844.03	27.38	1.29	1.17	-9.59
Wheat	VI	172.24	169.64	-1.51	615.46	91.10	-85.20	1282.93	740.29	-42.30	0.71	0.97	37.29
	VII	199.89	160.06	-19.92	454.82	99.35	-78.16	953.19	866.36	-9.11	1.00	0.87	-12.86
	VIII	86.76	150.48	73.44	891.84	117.52	-86.82	1671.60	770.91	-53.88	0.50	0.85	70.89
	Overall	161.14	158.44	-1.68	562.48	105.01	-81.33	1179.67	789.86	-33.04	0.78	0.89	14.20
Sugarcane (*suru*)	VI	193.15	305.56	58.20	60.52	128.63	112.53	85.04	62.25	-26.80	1.06	1.42	34.40
	VII	167.29	317.26	89.65	65.67	139.71	112.75	87.72	64.55	-26.41	1.05	1.37	30.50
	Overall	184.21	315.58	71.31	62.12	135.88	118.74	85.87	63.40	-26.17	1.06	1.40	31.80
Sugarcane (*Ratton*)	VI	269.78	258.81	-4.07	24.66	76.25	209.20	56.15	50.52	-10.03	1.59	1.76	10.69
	VII	220.74	308.40	39.71	29.12	78.25	168.71	59.87	52.68	-12.01	1.54	1.69	9.54
	Overall	242.27	290.82	20.04	26.75	77.48	189.63	57.75	52.01	-9.94	1.56	1.71	9.55

(Figures during the year 1980-81 are at constant prices)

Table 4.6 : Value of Input Saved per unit of Output due to New Crop Technology

Crop	*Zone*	*Rn (Rs.)*	*Ro (Rs.)*	*r*	*Sr(Rs.)*
Kharif sorghum	VII	591.01	891.62	0.34	1.99
	VIII	721.07	888.56	0.19	1.36
	Overall	654.04	891.45	0.27	1.74
Kharif paddy	II	561.04	1113.05	0.50	2.78
	V	520.37	1098.73	0.53	2.74
	IX	832.50	831.09	0.00	-0.01
	Overall	587.11	1014.29	0.42	2.47
Pearlmillet	V	731.91	2082.39	0.65	4.75
	VI	720.60	1474.35	0.51	3.68
	VII	821.84	1077.19	0.24	1.95
	Overall	745.62	1655.11	0.55	4.10
Groundnut	V	1426.82	1896.55	0.25	3.53
	VII	1821.84	2849.33	0.36	6.57
	Overall	1584.18	2243.01	0.29	4.65
Cotton	VII	1951.95	2727.03	0.28	5.55
	VIII	2482.97	2801.81	0.11	2.83
	Overall	2244.88	2477.80	0.09	2.11
Rabi sorghum	VI	807.00	736.17	-0.10	-0.78
	VII	887.77	580.50	-0.53	-4.70
	Overall	844.03	662.60	-0.27	-2.31
Wheat	VI	740.29	1282.93	0.42	3.13
	VII	866.36	953.19	0.09	0.79
	VIII	770.91	1671.6	0.54	4.15
	Overall	789.86	1179.67	0.33	2.61
Sugarcane (Suru)	VI	62.25	85.04	0.27	0.17
	VII	64.55	87.72	0.26	0.17
	Overall	63.40	85.87	0.26	0.17
Sugarcane (Ratton)	VI	50.52	56.12	0.10	0.05
	VII	52.68	59.87	0.12	0.06
	Overall	52.01	57.75	0.10	0.05

An alternative way of evaluation of the returns due to new technology is to estimate the quantity of extra output obtained with new technology using the old technology volume of inputs. The following expressions for this purpose were formulated.

Let, change in output per hectare in quintals (y) = Yn - Yo
r = the same as defined earlier,
(y) (r) = quantity of extra output due to technical change alone.

The results obtained with this approach are presented in Table 4.7. It can be seen that, during the year 2000-01, the per hectare quantity of extra output accrued with no additional cost to the farmers on resources came to 2.56 quintals in *kharif* sorghum, 4.00 quintals in *kharif* paddy, 3.92 quintals in pearlmillet, 0.96 quintal in groundnut, 0.44 quintal in cotton, 1.48 quintals in wheat, 19.66 quintals in *suru* sugarcane and 6.53 quintals in *ratoon* sugarcane. No impact of new technology was noticed in the case of *rabi* sorghum, as the quantity of extra output due to new technology was worked out to be negative.

The composite index of adoption of the crop production technology on the *rabi* sorghum farms, was also observed to be low (24.12 %), indicating the low level of technology adoption over a period of 20 years. This was the probable reason for lowering down the extra output of *rabi* sorghum due to new technology.

Table 4.7: Quantity of Extra Output per Hectare Due to Inputs with New Technology

Crop	*Zone*	*Yn (qtl.)*	*Yo (qtl.)*	*?*	*r*	*(?) (r)*	*% share of technology*
Kh. sorghum	VII	22.95	15.3	7.65	0.34	2.58	33.72
	VIII	16.95	7.11	9.84	0.19	1.85	18.85
	Overall	19.57	9.97	9.60	0.27	2.56	26.63
Kh. paddy	II	20.34	9.44	10.90	0.50	5.41	49.59
	V,	29.83	10.8	19.03	0.53	10.02	52.64
	IX	23.55	13.34	10.21	0.002	-0.02	-0.17
	Overall	21.25	11.75	9.50	0.42	4.00	42.12
Pearlmillet	V	16.00	2.6	13.40	0.65	8.69	64.85
	VI	9.46	4.21	5.25	0.51	2.68	51.12

(Contd.)

(Contd.Table 4.7)

Crop	*Zone*	*Yn (qtl.)*	*Yo (qtl.)*	*?*	*r*	*(?) (r)*	*% share of technology*
	VII	9.11	6.44	2.67	0.24	0.63	23.71
	Overall	11.07	3.93	7.14	0.55	3.92	54.95
Groundnut	V	18.68	13.26	5.42	0.25	1.34	24.77
	VII	8.00	5.66	2.34	0.36	0.84	36.06
	Overall	12.19	8.91	3.28	0.29	0.96	29.37
Cotton	VII	7.56	3.44	4.12	0.28	1.17	28.42
	VIII	5.48	3.63	1.85	0.11	0.21	11.38
	Overall	6.25	3.52	2.73	0.16	0.44	16.00
Rabi sorghum	VI	11.60	6.05	5.55	-0.10	-0.53	-9.62
	VII	9.60	8.67	0.93	-0.53	-0.49	-52.93
	Overall	10.55	7.06	3.49	-0.27	-0.96	-27.38
Wheat	VI	22.95	15.1	7.85	0.42	3.32	42.30
	VII	19.44	17.83	1.61	0.09	0.15	9.11
	VIII	16.54	8.08	8.46	0.54	4.56	53.88
	Overall	18.79	14.32	4.47	0.33	1.48	33.04
Sugarcane	VI	981.96	943.42	38.54	0.27	10.33	26.80
(suru)	VII	996.78	854.75	142.03	0.26	37.52	26.41
	Overall	989.17	914.03	75.14	0.26	19.66	26.17
Sugarcane	VI	795.29	768.76	26.53	0.10	2.65	9.98
(ratoon)	VII	781.16	718.91	62.25	0.12	7.48	12.01
	Overall	812.18	746.47	65.71	0.10	6.53	9.94

Amongst the agro-climatic zones, as like input saved approach, the quantity of extra output, due to new crop production technologies, was estimated in the range between 0.15 quintal in wheat in zone VII to 37.52 quintals in *suru* sugarcane in zone VII, during the period under study. In *kharif* paddy in zone IX and *rabi* sorghum in zones VI and VII, there was no extra output due to new technology.

The analysis of extra output due to technical change indicated that, the share of new technology in the increased per hectare output was found maximum in pearlmillet (54.95 %), whereas, it was the lower in *ratoon* sugarcane (9.94 %). This finding confirmed the impact of technology on the output of the crop enterprises. However, in the case of *rabi* sorghum, the influence of technology on the output was not observed. In other words, it can be said that, the increased productivity of this crop during the later period, was due to the increase in input use levels.

5

Changes in Resource Productivities and Allocation Efficiencies in the Production of Crops

The analysis presented in the previous chapter was related to the changes in the resource use structure, costs and returns of the selected crops grown by the sample farms in different agro-climatic zones in Maharashtra over a period of twenty years. There had been, however, some limitations of the type of analysis attempted, as it cannot measure the contribution of each specific resource in combination with other resources, responsible for changes in the output level during the specific time period as well as changes therein over a period of time. Moreover, productivity of individual resource undergoes a change because of complementarily and substitutability of resources alongwith new technology resulting from changes in resource use structure. The changes in resource use structure in turn indicate the improvements in allocation efficiency of the farms, as the later is dependent, besides prices of inputs and output, on productivities of resources associated with technological change in agriculture. The present chapter is, therefore, devoted to understand changes in the resource productivities and allocation efficiency in agriculture in different agro-climatic zones of Maharashtra during the period under reference by way of multivariate analysis of the production processes of the sample farms at two points of time.

The study of changes in resource productivities and allocation efficiency due to adoption of crop production technology is based on a comparison of the production functions estimated from the farm level data, for the years 1980-81 and 2000-01. Agro-climatic zonewise and cropwise comparison of production function was also attempted to study resource use efficiency in different crops. Further, production function estimates were deployed to estimate the returns to scale underlying agricultural production in the zones at two points of time and to suggest economic optima in the allocation of resources to maximize farm earnings. An attempt was also made at the end to find out the share of crop production technology in the productivity of individual crop enterprise, resulting from the technological change in agriculture over the period of 20 years.

5.1 The Production Function Model

The postulated production relationship in agriculture is reflected in the algebraic form of the function. 'Functions estimated from farm samples ordinarily have been of power form because of the smaller number of degrees of freedom involved in estimating the parameters, and partly because a multiplicative model has seemed logically appropriate' (Heady and Dillon, 1961).

The empirical evidences from previous studies suggest that amongst the many mathematical functions, Cobb-Douglas type of production function is the appropriate one for the study of resource productivities, because it specifies diminishing productivity and diminishing marginal rate of substitution among the factors and gives specific diminishing, increasing or constant returns. This function indicates that the coefficients of the ordinary least-square equations are the elasticities of production of the respective factors. The elasticities of production of inputs obtained from this function facilitate examination of resource use efficiency on neo-classical criterion that each factor of production is paid equal to its marginal productivity. This examination of efficiency / inefficiency in resource use is always relevant at the average level. The data were therefore, subjected to functional analysis by using the following form of Cobb-Douglas production function.

$$Y = a X_1^{b1} X_2^{b2} \ldots\ldots\ldots X_n^{bn} e^u$$

In this functional form, 'Y' is dependent variable, Xi's are independent resource variables, 'a' is constant representing intercept of the production function and bi's are the regression co-efficients of the respective resource variables. The regression co-efficients obtained from this function are also elastcities of production, which remain constant through relevant ranges of inputs. The sum of regression co-efficients i.e. bi's indicate the nature of returns to scale. When expressed in logarithmic terms, this function transforms into a linear form of the following type.

$$\text{Log } Y = \log a + b_1 \log X_1 + b_2 \log X_2 + \ldots\ldots + b_n \log X_n + u \log e$$

For fitting the production function, six inputs such as human labour, bullock power, seed, manure and fertilizers, other working capital and technological adoption index were considered as important factors in the production of crop. The output of the crop was used as dependent variable. The equation fitted was of the following form.

$$Y = a X_1^{b1} X_2^{b2} X_3^{b3} X_4^{b4} X_5^{b5} X_6^{b6} e^u$$

where, Y = Output of main produce (Qtl.) per hectare.
X_1 = Human labour (Mandays) per hectare.
X_2 = Bullock power (Pair-days) per hectare.
X_3 = Seed (Kg.) per hectare.
X_4 = Manure and fertilizers (Rs.) per hectare.
X_5 = Other working capital (Rs.) per hectare.
X_6 = Technology adoption index (%)
a = Intercept
bi's = Regression coefficients
u = Error term

The production functions with the above variables were estimated for individual crop separately for different agro-climatic zones and at the overall level, for the years 1980-81 and 2000-01. However, in the case of *ratoon* sugarcane, the seed input variable was excluded from the analysis.

In production function analysis, it is based on cross-sectional data, there often exists high correlation between some of the explanatory variables leading to the problem of multicollinearity. As a first step to ascertain the problem of multicollinearity, a zero-order correlation matrix for all the explanatory variables was obtained for each function. The following rule was applied to visualize the magnitude of multicollinearity. The correlation coefficients between a pair of explanatory variables is considered serious if it is greater than 0.80 (Heady and Dillon, 1961). In the present analysis, the major variables influencing the output of crop were confirmed with the help of correlation matrix technique. The values of different variables ranged from 0.0876 to 0.8157 in the case of all production functions. The problem of multicollinearity is serious if the inter-correlation between different factors is very high. There was thus, not serious problem of multicollinearity due to high inter-correlation between any two or more inputs. In case of majority of the pairs of explanatory variables, the correlation coefficient was less than 0.80. In case where correlation coefficient was greater than 0.80, also multicollinearity was not a problem, as the correlation coefficient corresponding to them was not high relative to the overall degree of multiple correlation among all the variables simultaneously (Klein, 1953), thus satisfying non-seriousness of the problem of multicollinearity in the analysis.

Use of different statistical tests of significance was also made to judge efficiency of the individual regression coefficients and coefficients of multiple correlation in respect of all the production functions. Further, returns to scale and difference between marginal value product of inputs and their prices have also been tested statistically. The procedure adopted for estimating different statistical tests of significance is already mentioned in Chapter 2. The factor share of the technology component in productivity of the crops was estimated by the decomposition analysis. Lastly, using Sustainability Index, the economic sustainability of crop production technology during the period under study was assessed. The results thus, obtained are discussed below.

5.2 Technological Change and Resource Productivity

As mentioned elsewhere, the coefficient of multiple determination (i.e. R^2) indicates the proportion of total variation in the dependent variable (i.e. crop output) explained by the independent resource variables jointly. At the same time, the regression coefficient of the individual resource variable is the production elasticity of respective resource in the Cobb-Douglas production function framework, which indicates the percentage change in crop output associated with a percentage change in the concerned input at its geometric mean level. On the whole, these results are of paramount importance as they provide readily the information relative to probable effects of resource use changes on crop output. While discussing the results, the cropwise inter-zonal comparisons are attempted in order to get idea regarding the variations in the productivities of different resources. Simultaneously, the resource productivities of individual crop for the two time periods, are also discussed in brief. The estimated parameters of the production functions of the selected crops are presented in Table 5.1 to Table 5.9, for the year 1980-81 and 2000-01, according to different agro-climatic zones and at the overall level of the zones.

***Kharif* sorghum :** The analysis of variance indicated the overall significance of all the production functions estimated for *kharif* sorghum.

Table 5.1 : Results of Cobb-Douglas Production Function of *kharif* sorghum in Different Agro-Climatic Zones

Particulars	*1980-81*			*2000-01*		
	VII	*VIII*	*Overall*	*VII*	*VIII*	*Overall*
Constant (a)	-0.7100	-0.4977	-1.2271	0.1218	-0.2388	0.7284
Human labour in	0.7916**	0.8344**	1.1394***	0.3151	0.5215*	0.3010
Mandays (X1)	(0.3821)	(0.4252)	(0.2160)	(0.3806)	(0.3696)	(0.2265)
Bullock power in	-0.0701	0.3540	0.3000	0.0344	0.1150	0.2485***
Pairdays (X2)	(0.2980)	(0.3171)	(0.2039)	(0.0986)	(0.1218)	(0.0900)
Seed in Kg. (X3)	-0.1086	-0.5600	-0.1832	0.2451	-0.0878	-0.8397**
	(0.4796)	(0.6767)	(0.3939)	(0.6348)	(0.8865)	(0.4441)
Manures and fertilizers	0.0663	-0.0051	0.0201	0.1464***	-0.0056	0.1233**
in Rs. (X4)	(0.0883)	(0.0521)	(0.0399)	(0.0615)	(0.1057)	(0.0508)
Other working capital	-0.0466	-0.2346**	-0.1183***	0.0373	0.2882***	0.1166*
in Rs. (X5)	(0.0743)	(0.1054)	(0.0457)	(0.1421)	(0.1055)	(0.0718)
Technology adoption	0.3005*	0.2441	0.2403**	-0.1049	-0.2821	0.5215**
index in Per cent (X6)	(0.2060)	(0.2123)	(0.1227)	(0.3127)	(0.4133)	(0.2425)
R2	0.6501	0.6399	0.8092	0.5755	0.5368	0.5344
F value	6.69 ***	6.04 ***	29.69 ***	4.88 ***	4.17 ***	8.03 ***

(Figures in parentheses indicate the standard errors of the respective coefficients.)

Note :- *, ** and *** indicate the level of significance at 10, 5 and 1 per cent.

The value of R^2 (coefficient of multiple determination) turned out to be quite high and was statistically significant beyond 1 per cent level in both the years. The included variables explained 80.92 and 53.44 per cent of the variations in the logarithms of the output in the year 1980-81 and 2000-01, respectively. The regression coefficients of human labour, and technology adoption were positive and highly significant, thereby indicated that these resource variables have positive influence on the output of the crop, whereas, the regression coefficient of other working capital was negative and highly significant in the year 1980-81. While, the regression coefficients of bullock power, manures and fertilizers, other working capital and technology adoption index were found positive and significant in the year 2000-01. The regression coefficient of seed has turned out to negative and significant, indicating the excess use of this resource in the production of *kharif* sorghum, at the overall level.

A critical examination of the production elasticities of the individual input across the zones, revealed that the regression coefficients of human labour in zones VII and VIII and technology adoption index in zone VII were found positive and highly significant, whereas, the regression coefficient of the working capital in zone VIII turned out negative and significant, during the year 1980-81. The situation was changed during the year 2000-01 and the regression coefficients of manure and fertilizers in zone VII zone and the regression coefficients of human labour and other working capital in zone VIII turned positive and highly significant. The significant and positive coefficients clearly indicate that these variables exert significant influence on the output of the *kharif* sorghum.

***Kharif* paddy** : The 'F' ratio obtained from the analysis of variance was found to be highly significant indicating thereby overall significance of the estimated production functions. The six resource variables included in the production function analysis have explained jointly a little over 51 per cent variation during both the periods in the paddy output at the overall level. Among the agro-climatic zones, the proportion of total variation explained by the concerned resource variables, was ranged between 23.42 and 79.56 per cent in the year 1980-81, whereas, it was ranged between 69.33 and 85.25 per cent in 2000-01. The lowest and the highest proportions being in IX zone and II zone during the year 1980-81, and this trend was reverse during the year 2000-01.

The regression coefficients of seed, manure and fertilizers and technology adoption index during the year 1980-81 and the regression coefficients of all the resource variables, excepting bullock power, during the year 2000-01 have turned out to be positive and significant. However, that

Table 5.2 : Results of Cobb-Douglas Production Function of *Kharif* Paddy in Different Agro-Climatic Zones

Particulars	*1980-81*				*2000-01*			
	II	*V*	*IX*	*Overall*	*II*	*V*	*IX*	*Overall*
Constant (a)	-1.7199	-1.7841	-0.4773	-3.3845	5.0166	-1.3980	-2.5191	0.1615
Human labour in Mandays (X1)	0.1097	-0.2285	0.1523	0.2614	0.1968	0.3929*	0.4961***	0.1610*
	(0.5510)	(0.2511)	(0.5343)	(0.2135)	(0.2160)	(0.2721)	(0.1381)	(0.1004)
Bullock power in Pairdays (X2)	1.4148	0.1358	0.2199	0.0073	-0.2677***	0.1136	0.1751**	-0.0318
	(1.0789)	(0.2311)	(0.4421)	(0.2519)	(0.0980)	(0.1284)	(0.0922)	(0.0648)
Seed in Kg. (X3)	-0.2180	1.3107**	0.5620	1.6274***	-2.0444*	0.9550*	1.2185**	0.6004*
	(1.1218)	(0.6880)	(1.1199)	(0.5786)	(1.2799)	(0.6431)	(0.5855)	(0.4226)
Manures and fertilizers in Rs. (X4)	0.3156**	0.2962***	0.1090	0.1844***	-0.1460	0.1245	0.2699**	0.2190**
	(0.1467)	(0.0845)	(0.0900)	(0.0649)	(0.1846)	(0.1771)	(0.1323)	(0.1051)
Other working capital in Rs. (X5)	-0.0269	0.0073	0.1079	-0.1292**	0.0962*	0.2039**	0.2223***	0.2226***
	(0.1208)	(0.1073)	(0.1448)	(0.0554)	(0.0702)	(0.1054)	(0.0690)	(0.0398)
Technology adoption index in Per cent (X6)	0.6011**	0.2117	0.0069	1.0334***	0.2810	-0.5463	-0.7179*	1.1311***
	(0.2814)	(0.2639)	(0.4242)	(0.3216)	(0.5693)	(0.6423)	(0.4681)	(0.3512)
R2	0.7956	0.6877	0.2342	0.5129	0.6933	0.7551	0.8525	0.5199
F value	10.90 ***	6.17 ***	0.86	12.64 ***	6.33 ***	8.63 ***	16.18 ***	12.99***

(Figures in parentheses indicate the standard errors of the respective coefficients.)

Note :- *, ** and *** indicate the level of significance at 10, 5 and 1 per cent.

of other working capital during the year 1980-81, was found negative in the case of production function estimated, at the overall level. Only two resource variables viz. seed and technology adoption index during the year 1980-81, at the overall level, and technology adoption index during the year 2000-01 had production elasticity exceeding the unity, indicating that there was a tremendous potential of increasing production through the allocation of these two variables to a greater extent in favour of paddy.

Among the agro-climatic zones, during the year 1980-81, the regression coefficients of manure and fertilizers, and technology adoption index in zone II and that of seed and, manure and fertilizers in zone V were positive and significant. None of the resource variables found significant in zone IX. The picture was quite different during 2000-01, as the regression coefficients of other working capital in zone II, human labour, seed and other working capital in zone V and all the resource variables, except technology adoption index in zone IX were positive and significant. The regression coefficients of bullock power and seed in zone II and technology adoption index in zone IX were turned out to negatively significant, indicating their excess use in paddy production.

Pearlmillet : The overall significance of the estimated production function have been indicated by high significant value of the 'F' ratios obtained from the analysis of variance in case of pearlmillet production. The R^2 was 47.37 and 72.65 per cent at the overall level, during the year 1980-81 and 2000-01, respectively. This indicated the magnitude of the total variation explained by the six variables jointly. The regression coefficients of human labour and bullock power have turned out to positive and significant during the year 1980-81, whereas, the regression coefficients of human labour, bullock power, manure and fertilizers, and other working capital were positively significant during the year 2000-01, at the overall level.

The examination of agro-climatic zonewise results of pearlmillet production functions revealed that during the year 1980-81, the total variation in the output of pearlmillet explained jointly by the concerned resource variables was the highest in zone VI (82.16 %) and the lowest was in zone VII (42.81 %), whereas, during the year 2000-01, the highest and lowest were in zone VI (90.11 %) and in zone V (76.88 %), respectively. The regression coefficient of human labour in all zones, bullock power in zone VI and technology adoption index in zone VII were the important resource variables contributing to paddy output. The regression coefficients of manure and fertilizers in zones VI and VII, and other working capital in zone VII were negatively significant indicating the excess use during the year

Table 5.3 : Results of Cobb-Douglas Production Function of Pearlmillet in Different Agro-Climatic Zones

Particulars	*1980-81*				*2000-01*			
	V	*VI*	*VII*	*Overall*	*V*	*VI*	*VII*	*Overall*
Constant (a)	-1.0533	-4.5220	-0.1032	-2.7774	0.5627	-1.0942	-0.8893	0.3335
Human labour in	0.8714**	1.2433*	0.3666**	0.9482***	0.2705*	0.7458***	0.3744	0.4687***
Mandays (X1)	(0.3692)	(0.8052)	(0.1838)	(0.2842)	(0.1620)	(0.1103)	(0.2972)	(0.0793)
Bullock power in	0.0857	2.4069***	0.3006	1.5552***	0.1381	0.0490	-0.0584	0.1063***
Pairdays (X2)	(0.8869)	(0.8897)	(0.2289)	(0.5427)	(0.1989)	(0.0405)	(0.4096)	(0.0424)
Seed in Kg. (X3)	-0.1397	0.3053	-0.0612	0.4129	-0.3876	-0.4705	-0.1729	-0.0911
	(0.6134)	(0.4885)	(0.1164)	(0.4219)	(0.4675)	(0.3804)	(1.0844)	(0.2864)
Manures and fertilizers	-0.0885	-0.1534**	-0.0292*	-0.0090	0.0479*	-0.0575	-0.0292	0.0634**
in Rs. (X4)	(0.0820)	(0.0634)	(0.0198)	(0.0444)	(0.0339)	(0.0338)	(0.0547)	(0.0230)
Other working capital	-0.0644	0.1492	-0.0641*	-0.0616	-0.0258	0.2293**	0.2400*	0.1498***
in Rs. (X5)	(0.1387)	(0.1171)	(0.0410)	(0.0919)	(0.1200)	(0.0874)	(0.1555)	(0.0602)
Technology adoption	0.2697	0.1729	0.1486**	-0.0166	0.1013	0.4340*	0.5737	0.0897
index in Per cent (X6)	(0.2025)	(0.2706)	(0.0830)	(0.3051)	(0.2338)	(0.2597)	(0.5864)	(0.1577)
R2	0.4792	0.8216	0.4281	0.4737	0.7688	0.9011	0.7796	0.7265
F value	2.58	12.89 ***	1.20	9.72 ***	9.31 ***	25.51 ***	5.66 ***	28.69 ***

(Figures in parentheses indicate the standard errors of the respective coefficients.)

Note :- *, ** and *** indicate the level of significance at 10, 5 and 1 per cent.

1980-81. The regression coefficients of human labour in zones V and VI, manure and fertilizers in zone V, other working capital in zones VI and VII, and technology adoption index in zone VI were positive and significant, during the year 2000-01. None of the resource variables had production elasticity exceeding unity, during the later period.

Groundnut : The production functions estimated for groundnut crop have been observed to be highly significant. At the overall level, 85.57 and 81 per cent of the total variation in the output of groundnut have been explained jointly by the resource variables included in the production functions. The regression coefficients of human labour, bullock power and technology adoption index were positive and significant during both the years, whereas, the regression coefficients of manure and fertilizers during the year 2000-01 were positive and significant. The regression coefficients of seed and other working capital during the year 1980-81 were negatively significant, while the same were turned out to negative but non-significant during the later year.

Concerning the agro-climatic zones, during the year 1980-81 the regression coefficients of bullock power, seed and technology adoption index were positive and significant in zone V, while, the regression coefficients of human labour and technology adoption index were positively significant in zone VII. However, the regression coefficients of other working capital in both the zones and that of bullock power in zone VII were turned out to negative and significant. It is to note that the regression coefficient of other working capital found negatively significant indicating the inefficient use of the capital in the production of groundnut during the year 1980-81. In the year 2000-01, the regression coefficients of human labour and, manure and fertilizers were found positive and significant in both the zones under study.

Cotton : The 'F' ratio obtained from the production analysis have been turned out to highly significant, indicating thereby overall significance of the estimated production functions in respect of cotton crop. The proportion of total variation in the output of cotton crop explained jointly by the resource variables was 67.10 and 44.50 per cent at the overall level, during the year 1980-81 and 2000-01, respectively. The regression coefficients of human labour and, manure and fertilizers were positive and significant, whereas, that of seed was negatively significant during the year 1980-81. While, during the year 2000-01, the regression coefficients of human labour and technology adoption index were positive and significant at the overall level.

Table 5.4 : Results of Cobb-Douglas Production Function of Groundnut in Different Agro-Climatic Zones

Particulars	*1980-81*			*2000-01*		
	VII	*VIII*	*Overall*	*VII*	*VIII*	*Overall*
Constant (a)	-8.4126	3.3566	-4.1976	0.2651	1.5448	0.5927
Human labour in	-0.7090	3.1165***	2.7421***	0.4011*	0.7430***	1.0179***
Mandays (X1)	(0.8164	(1.0836)	(0.7382)	(0.2476)	(0.1909)	(0.1354)
Bullock power in	1.3001***	-1.4293*	1.8978***	0.0326	0.1665	0.4002***
Pairdays (X2)	(0.4674)	(0.9731)	(0.4453)	(0.1892)	(0.2769)	(0.1344)
Seed in Kg. (X3)	4.3527***	-3.6602	-1.7637*	-0.2363	-0.9393	-0.3253
	(1.3837)	(3.2995)	(1.2130)	(0.3709)	(1.3318)	(0.4598)
Manures and fertilizers	-0.0987	0.0620	0.0502	0.0971***	0.1149*	0.1064***
in Rs. (X4)	(0.0909)	(0.1116)	(0.0872)	(0.0304)	(0.0808)	(0.0371)
Other working capital	-0.2586***	-0.2504**	-0.2257***	0.0408	-0.0610	-0.0148
in Rs. (X5)	(0.0640)	(0.1173)	(0.0757)	(0.0548)	(0.1053)	(0.0521)
Technology adoption	0. 7399***	0.5229**	0.5593*	0.1285	-0.2409	0.6173**
index in Per cent (X6)	(0.2320)	(0.2937)	(0.3498)	(0.1432)	(0.4860)	(0.2779)
R2	0.9162	0.8973	0.8557	0.857	0.6571	0.8100
F value	30.61 ***	24.46 ***	42.70 ***	16.78 ***	5.37 ***	30.69 ***

(Figures in parentheses indicate the standard errors of the respective coefficients.)

Note :- *, ** and *** indicate the level of significance at 10, 5 and 1 per cent.

Table 5.5 : Results of Cobb-Douglas Production Function of Cotton in Different Agro-Climatic Zones

Particulars	*1980-81*			*2000-01*		
	VII	*VIII*	*Overall*	*VII*	*VIII*	*Overall*
Constant (a)	-1.3704	-0.6705	-1.0623	0.2355	-0.9951	0.7872
Human labour in	0.8689**	0.6310*	0.6909**	0.0816	-0.0662	0.4033**
Mandays (X1)	(0.3678)	(0.4308)	(0.2883)	(0.5399)	(0.1891)	(0.2389)
Bullock power in	-0.0857	0.3072	0.2206	-0.0361	0.7507***	0.0804
Pairdays (X2)	(0.5320)	(0.4153)	(0.3093)	(0.2394)	(0.1828)	(0.0759)
Seed in Kg. (X3)	-0.1998	-0.1387	-0.2155**	0.2620*	-0.1282*	-0.0454
	(0.1648)	(0.1387)	(0.0962)	(0.1709)	(0.0942)	(0.1185)
Manures and fertilizers	0.0899*	0.0814**	0.0544*	0.0411	-0.0831	0.0690
in Rs. (X4)	(0.0658)	(0.0441)	(0.0361)	(0.0818)	(0.1187)	(0.0708)
Other working capital	-0.0333	0.1408**	0.0015	-0.0597	0.0435	0.0306
in Rs. (X5)	(0.0793)	(0.0649)	(0.0521)	(0.1284)	(0.0638)	(0.0723)
Technology adoption	0.3444	-0.3806	0.1312	0.2608	0.8035*	0.5132*
index in Per cent (X6)	(0.4970)	(0.2848)	(0.2064)	(0.4342)	(0.4808)	(0.3637)
R2	0.7369	0.8281	0.6710	0.4543	0.8961	0.4450
F value	7.28 ***	13.49 ***	13.46 ***	2.16	24.15 ***	5.29 ***

(Figures in parentheses indicate the standard errors of the respective coefficients.)

Note :- *, ** and *** indicate the level of significance at 10, 5 and 1 per cent.

Between two zones, during the both the time periods, i.e. 1980-81 and 2000-01, the resource variables have explained variation at the highest in zone VIII (82.81 % and 89.61 %) and the lowest in zone VII (73.69 % and 45.43 %), respectively. In both the zones, during the year 1980-81, the regression coefficients of human labour and, manure and fertilizers were positive and significant and their magnitude was more in zone VII. The regression coefficient of other working capital was found positive and significant in zone VIII, while it was turned out negative but non-significant in zone VII. In the year 2000-01, the regression coefficient of seed in zone VII and that of bullock power and technology adoption index in zone VIII were positive and significant. The regression coefficients of seed, however, turned out to negative and significant in zone VIII, indicating no proper use of seeds.

***Rabi* sorghum :** The estimated production function for *rabi* sorghum crop have been observed to be significant as the 'F' value obtained from analysis of variance were highly significant. The resource variables included in the production function explained 87 and 74 per cent of the total variation in the output of *rabi* sorghum crop at the overall level, during the year 1980-81 and 2000-01, respectively. The regression coefficients of bullock power, manure and fertilizers, and technology adoption index were positively significant during the year 1980-81, while the regression coefficients of human labour, seed, other working capital and technology adoption index were positive and significant during the year 2000-01. The regression coefficients of other working capital during the year 1980-81 and that of bullock power during the year 2000-01 were, however, negatively significant, at the overall level, indicating negative influence on output.

Amongst the zones, the proportion of total variation explained jointly by the six resource variables was the highest in zone VI (95.93 % and 81.84 %) during both the years, respectively. During the year 1980-81, the regression coefficients of bullock power and technology adoption index in zone VI and, manure and fertilizers in zone VII were found positive and significant, whereas that of for other working capital in zone VI and seed in zone VII was negative and significant. In the year 2000-01, the situation was quite changed as the regression coefficients of seed and manure and fertilizers in zone VI and that of human labour in zone VII were turned out to positive and significant. None of the resource variables, excepting seed in zone VI during the year 2000-01, had production elasticity exceeding unity.

Wheat : The 'F' ratio from the analysis of variance have turn out to be highly significant, during both the time periods at the overall level. The

Table 5.6 : Results of Cobb-Douglas Production Function of *Rabi* Sorghum in Different Agro-Climatic Zones

Particulars	*1980-81*			*2000-01*		
	VI	*VII*	*Overall*	*VI*	*VII*	*Overall*
Constant (a)	-1.3704	-0.6705	-1.0623	0.2355	-0.9951	0.7872
Constant (a)	-0.5242	0.6312	-1.7454	-0.469	-0.8317	1.2310
Human labour in	0.0927	0.2117	0.1848	0.1565	0.4109***	0.1832**
Mandays (X1)	(0.1251)	(0.3622)	(0.1754)	(0.1665)	(0.1453)	(0.0826)
Bullock power in	0.4644**	0.5416	0.9913***	-0.0216	-0.0179	-0.1558*
Pairdays (X2)	(0.2423)	(0.5152)	(0.1831)	(0.1524)	(0.1973)	(0.0969)
Seed in Kg. (X3)	-0.1004	-0.5756**	0.0772	1.3702**	0.0655	1.4830***
	(0.3579)	(0.3060)	(0.2410)	(0.6092)	(0.9202)	(0.4458)
Manures and fertilizers	0.0180	0.1311***	0.0357**	0.1614**	-0.9520	0.0563
in Rs. (X4)	(0.0144)	(0.0281)	(0.0180)	(0.0822)	(0.0930)	(0.0510)
Other working capital-	0.0689**	-0.0234	-0.1338***	0.0569	0.2053	0.1509*
in Rs. (X5)	(0.0362)	(0.0670)	(0.0442)	(0.1181)	(0.2302)	(0.0938)
Technology adoption	0.7452***	-0.0192	0.8619***	-0.4835	0.4394	0.8495**
index in Per cent (X6)	(0.0707)	(0.4119)	(0.1496)	(0.4488)	(0.6053)	(0.3254)
R	0.9593	0.9016	0.8685	0.8184	0.7887	0.7384
F value	66.00 ***	18.33 ***	39.63 ***	12.62 ***	7.47 ***	16.94 ***

(Figures in parentheses indicate the standard errors of the respective coefficients.)

Note :- *, ** and *** indicate the level of significance at 10, 5 and 1 per cent.

proportion of total variations explained jointly by the resource variables was 49.55 and 64.80 per cent of the total variations in 1980-81 and 2000-01, respectively. The regression coefficients of manure and fertilizers, and other working capital during the year 1980-81 and that of for human labour, manure and fertilizers, other working capital and technology adoption index during the year 2000-01, were positive and significant. These variables influenced the yield. The regression coefficient of seed during the year 1980-81 was, however, turned out to negative and significant with the high magnitude, indicating excessive use or misuse of seeds.

Among the three zones, i.e. zones VI, VII and VIII, the proportion of total variation in the output of wheat crop explained jointly by the concerned resources variables ranged between 32.82 and 82.63 per cent during the year 1980-81, whereas, the proportion for the same ranged between 65 and 85.14 per cent during the year 2000-01. The lowest and the highest proportions being in zones VIII and VI during 1980-81 and in zones VI and VII during the year 2000-01, respectively. During the year 1980-81, the regression coefficients of human labour and other working capital in zone VI, the regression coefficient of technology adoption index in zones VI and VII, and the regression coefficient of manure and fertilizers in zone VIII were positive and significant. While, the regression coefficient of manure and fertilizers in zone VI and the regression coefficient of bullock power in zone VIII were turned out to negatively significant. The picture was changed during the year 2000-01. The regression coefficients of other working capital and technology adoption index in zone VI, the regression coefficient of the technology adoption index only in zone VII and the regression coefficients of human labour, bullock power, and other working capital in zone VIII were turned out to positive and significant.

***Suru* sugarcane :** The production function estimated for *suru* sugarcane were highly significant as the 'F' ratios obtained from the analysis of variance were high. At the overall level, the six resource variables explained jointly 92.77 and 36.27 per cent of the total variation in the output of the *suru* sugarcane. The regression coefficients of human labour, manure and fertilizers, and technology adoption index during the year 1980-81 were positive and significant. The regression coefficients of human labour and seed, however, were turned out to negative and significant during the year 2000-01. The regression coefficients of manure and fertilizers, and technology adoption index have contributed significantly in the output of the *suru* sugarcane during the year 2000-01.

With regards to the zones, the proportion of total variation explained jointly by the independent variables ranged between 87.87 and 96.54 per

Table 5.7 : Results of Cobb-Douglas Production Function of Wheat in Different Agro-Climatic Zones

Particulars	*1980-81*				*2000-01*			
	VI	*VII*	*VIII*	*Overall*	*VI*	*VII*	*VIII*	*Overall*
Constant (a)	-1.7090	-5.7689	-0.8733	6.8321	0.6024	0.6446	-1.6459	-1.0899
Human labour in	1.0836***	1.8002	0.1607	0.8188	0.1045	0.0814	0.2744*	0.3500***
Mandays (X1)	(0.3362)	(1.5390)	(2.5570)	(1.0717)	(0.1324)	(0.1651)	(0.1570)	(0.0817)
Bullock power in	-0.0005	-0.3895	-1.8457*	0.4461	0.0398	0.0395	0.2866*	0.0234
Pairdays (X2)	(0.1856)	(0.8231)	(1.1372)	(0.5025)	(0.0342)	(0.0638)	(0.1692)	(0.0237)
Seed in Kg. (X3)	-0.0612	1.5733	0.8817	-3.5470***	-0.4032	-0.5399	0.1918	-0.0307
	(0.7881)	(2.3183)	(2.4331)	(1.0438)	(0.7340)	(0.5276)	(0.3030)	(0.2652)
Manures and fertilizers	-0.0758**	-0.0306	0.1583**	0.1886**	-0.0434	-0.0212	0.0980	0.2327**
in Rs. (X4)	(0.0417)	(0.1574)	(0.0888)	(0.0521)	(0.0486)	(0.0925)	(0.2971)	(0.0329)
Other working capital	0.3287***	0.0427	0.2139	0.4405*	0.2817**	0.1857	0.2706***	0.2592***
in Rs. (X5)	(0.0785)	(0.1744)	(0.3878)	(0.1824)	(0.1478)	(0.1843)	(0.0900)	(0.0759)
Technology adoption	0.2172***	0.8038*	0.6093	0.5251	0.3265*	0.6386*	0.1325	0.5770*
index in Per cent (X6)	(0.0794)	(0.4691)	(0.6783)	(0.4799)	(0.2098)	(0.4606)	(0.6449)	(0.2357)
R2	0.8263	0.4852	0.3282	0.4955	0.6500	0.8100	0.8514	0.6480
F value	11.42 ***	2.26	1.27	9.63 ***	4.46 ***	10.23 ***	14.90 ***	18.04 ***

(Figures in parentheses indicate the standard errors of the respective coefficients.)

Note :- *, ** and *** indicate the level of significance at 10, 5 and 1 per cent.

cent during the year 1980-81, while the proportion for the same were ranged between 51.06 and 82.39 per cent during the year 2000-01. The highest proportions were noticed in zone VII during both the years i.e. 1980-81 and 2000-01. The regression coefficients of human labour, manure and fertilizers, and technology adoption index in zone VI, the regression coefficient of manure and fertilizers in zone VII were positive and significant during the year 1980-81. In the year 2000-01 it was found that, the regression coefficients of human labour and seed in zone VI and the regression coefficients of seed and, manure and fertilizers in zone VII were turned out to negative and significant, indicating thereby their excess use in *suru* sugarcane production. The regression coefficients of other working capital and technology adoption index were found positive and significant in zone VII during the year 2000-01. None of the resource variables exceeds the unity during both the time periods in *suru* sugarcane production.

***Ratoon* sugarcane :** The five resource variables included in the production function analysis of *ratoon* sugarcane have explained jointly 85.23 and 50.13 per cent of the total variation at the overall level, during the year 1980-81 and 2000-01, respectively. At the overall level, the regression coefficients of human labour, manure and fertilizers, and technology adoption index were positive and significant during the year 1980-81, whereas, the regression coefficients of human labour and, manure and fertilizers were turned out to positive and significant during the year 2000-01.

Concerning to the agro-climatic zones, the proportion of total variation explained by the independent variables ranged between 78.81 and 91.27 per cent during the year 1980-81, the proportion for the same ranged between 44.71 and 70.09 per cent in the year 2000-01. The regression coefficients of manure and fertilizers, and technology adoption index were positive and significant in both the zones under study. The regression coefficient of bullock power was turned out to negatively significant in zone VI. It is worthwhile to note that the regression coefficient of technology adoption index exceeds the unity in zone VI during the year 1980-81. In the year 2000-01, it was observed that, the regression coefficient of human labour was positive and significant in both the zones. The regression co-efficient of other working capital was positive and significant in zone VI, while, it was found negatively significant in zone VII.

In general, it was observed that the production elasticities of human labour, bullock power and seed inputs have declined, while, those of manure and fertilizers, other working capital and technology adoption index increased to some extent during the period. The increase in the magnitudes of the production elasticities of these resource variables during the

Table 5.8 : Results of Cobb-Douglas Production Function of *suru* Sugarcane in Different Agro-Climatic Zones

Particulars	*1980-81*			*2000-01*		
	VI	*VII*	*Overall*	*VI*	*VII*	*Overall*
Constant (a)	3.9557	-1.5854	3.0442	4.3281	5.9138	4.4973
Human labour in	0.2675*	0.1125	0.2664**	-0.2883***	-0.0392	-0.1504**
Mandays (X1)	(0.1741)	(0.3460)	(0.1018)	(0.1222)	(0.1022)	(0.0855)
Bullock power in	0.0021	0.0417	0.0253	0.0354	0.0119	0.0179
Pairdays (X2)	(0.0630)	(0.0480)	(0.0188)	(0.0319)	(0.0222)	(0.0214)
Seed in Kg. (X3)	-0.0612	-0.0894	0.0234	-0.3248**	-0.6279***	-0.3459***
	(0.7881)	(0.0725)	(0.0237)	(0.1678)	(0.1695)	(0.1117)
Manures and fertilizers	0.1856**	0.2888**	0.2404***	-0.0162	-0.3646**	0.3145**
in Rs. (X4)	(0.1017)	(0.1661)	(0.0651)	(0.1082)	(0.1277)	(0.1798)
Other working capital	0.0164	0.7552	0.0040	0.0093	0.0925**	0.0253
in Rs. (X5)	(0.1029)	(0.6122)	(0.7950)	(0.0344)	(0.0424)	(0.0248)
Technology adoption	0.5452***	0.1222	0.4979***	0.3606	0.7813**	0.5128**
index in Per cent (X6)	(0.1732)	(0.4603)	(0.1219)	(0.2985)	(0.2794)	(0.1958)
R2	0.8787	0.9654	0.9277	0.5104	0.8239	0.3627
F value	20.28 ***	33.48 ***	56.46 ***	2.92 **	7.49 ***	2.96 **

(Figures in parentheses indicate the standard errors of the respective coefficients.)

Note :- *, ** and *** indicate the level of significance at 10, 5 and 1 per cent.

Table 5.9 : Results of Cobb-Douglas Production Function of *Ratoon* Sugarcane in Different Agro-Climatic Zones

Particulars	*1980-81*			*2000-01*		
	VI	*VII*	*Overall*	*VI*	*VII*	*Overall*
Constant (a)	5.5446	5.7135	4.9568	1.4601	2.1225	2.1106
Human labour in	0.0652	0.0105	0.2146*	0.2803**	0.3782***	0.1882***
Mandays (X1)	(0.1862)	(0.4683)	(0.1250)	(0.1200)	(0.0883)	(0.0659)
Bullock power in	-0.1533*	0.0434	0.0171	0.0108	0.0055	-0.0057
Pairdays (X2)	(0.0967)	(0.0535)	(0.0241)	(0.0268)	(0.0218)	(0.0150)
Manures and fertilizers	0.1486**	0.2977*	0.1143**	-0.0353	0.047	0.2465*
in Rs. (X3)	(0.0730)	(0.2201)	(0.0613)	(0.1222)	(0.0655)	(0.1588)
Other working capital	0.0157	-0.1061	-0.0275	0.1184*	-0.0792*	0.0259
in Rs. (X4)	(0.1262)	(0.2742)	(0.8940)	(0.0794)	(0.0451)	(0.0381)
Technology adoption	1.015***	0.8457**	0.8397***	0.3606	-0.0441	0.0093
index in Per cent (X5)	(0.1800)	(0.4135)	(0.1315)	(0.3707)	(0.1551)	(0.1313)
R2	0.9127	0.7881	0.8523	0.4471	0.7009	0.5013
F value	23.00 ***	8.18 ***	32.31 ***	1.78	5.16 ***	5.63 ***

(Figures in parentheses indicate the standard errors of the respective coefficients.)

Note :- *, ** and *** indicate the level of significance at 10, 5 and 1 per cent.

period supported the impact of the technological change during the period. The findings of this study confirmed the study results of Rahman and Rahman (1999). They indicated that the modern technology diffusion had contributed substantially to increase in crop production as well productivity, however, it had not spread uniformly across regions.

The decrease in the production elasticities in human labour, bullock power and seed clearly supports the hypothesis that the relative importance of human labour and bullock power would decline in the process of transformation of agriculture from traditional to modern methods. On the other hand, the increase in the production elasticities of manure and fertilizers, other working capital (which includes the expenditure on machine charges, irrigation, plant protection measures, land revenue and other taxes, repairs to the farm implements, incidental charges, etc.) and technology adoption index indicated that, there was a greater scope for intensifying applications of these inputs in agricultural production in the zones. Moreover, based on higher production elasticities of non-conventional inputs during the later period, it would be more correct to say that the process of transforming agriculture in different agro-climatic zones, places a higher premium on investment in technological change.

The production elasticities of all the input variables, with few exceptions, were less than unity in both the years, implying decreasing marginal productivity of factor inputs. This means that if other inputs are fixed at a specified level and only one input is increased, the gross returns will increase at a diminishing rate. This result was confirmed with the findings of Tiwari and Dheer (1994). They showed that the per hectare elasticities of production to be less than unity indicating the diminishing marginal returns for all factor of production.

The comparison of intercept term i.e. constant (Log_{10} a) revealed that there has been technological change in production of majority crop enterprises, except in wheat and *ratoon* sugarcane, as the magnitude of intercept term was higher during the later year over the base year. The degree of technological change, however, varied considerably among the crops as well as across the agro-climatic zones.

The estimated production functions indicated on an average, the type of changes in resource productivities that have been experienced in agriculture in different agro-climatic zones of Maharashtra during the period under consideration. These estimates, though valid on an average, might tend to iron out and thereby conceal some of the differences, which may prevail between sample farms of different agro-climatic zones. Many a times, a resource productivity and resource use efficiency are greatly influenced by the farm size. Besides this, the rate of adoption of new technology is dependent, to some extent, on farm size as the later also reflects the ability of farms in owing the resources required for optimization of the farm returns. On the other hand, in the absence of the alternative employment opportunities on the farm, the use of labour input was excessive in some crops.

5.3 Changes in Returns to Scale

As mentioned earlier, the regression coefficients in the Cobb-Douglas production functions are the production elasticities and their sum indicates the returns to scale. The returns to scale are increasing, constant, or decreasing according as the sum of regression coefficients is greater than, equal to, or less than unity. Accordingly, the sum of production elasticities of all the resource variables have been worked out for the individual crop production functions for all the crops in different agro-climatic zones for the year 1980-81 and 2000-01. All the sums of production elasticities have been tested statistically for their deviation from the unity with the help of 'Z' test. The results indicating the returns to scale on sample farms, during both the periods are presented in Table 5.10. The sum of bi's do not statistically differ from unity as indicated by Z test and therefore, there existed a constant returns to scale

Table 5.10 : Returns to Scale for Major Crops in Different Agro-Climatic Zones

Crop	*Zone*	*Sum of bi's in 1980-81*	*Returns to scale indicated by Z* test*	*Sum of bi's in 2000-01*	*Returns to scale indicated by Z* test*
Kharif sorghum	VII	0.9331	Constant	0.6734	Constant
	VIII	0.6328	Constant	0.5492	Constant
	Overall	1.3983	Constant	0.4692	Constant
Kharif paddy	II	2.1914	Constant	-1.8841	Constant
	V	1.7332	Constant	1.2436	Constant
	IX	1.1580	Constant	1.6640	Constant
	Overall	2.9847	Constant	2.3023	Constant
Pearlmillet	V	0.9342	Constant	0.1444	Constant
	VI	4.1242	Constant	0.9301	Constant
	VII	0.6613	Constant	0.9276	Constant
	Overall	2.8291	Constant	0.7868	Constant
Groundnut	V	5.3264	Constant	0.4638	Constant
	VII	-1.6385	Constant	-0.2168	Constant
	Overall	3.2600	Constant	1.8017	Constant
Cotton	VII	0.9844	Constant	0.5497	Constant

(Contd.)

(Contd.Table)

Crop	Zone	Sum of bi's in 1980-81	Returns to scale indicated by Z* test	Sum of bi's in 2000-01	Returns to scale indicated by Z* test
	VIII	0.6411	Constant	1.3202	Constant
	Overall	0.8831	Constant	1.0511	Constant
Rabi sorghum	VI	1.1510	Constant	1.2399	Constant
	VII	0.2662	Constant	0.1512	Constant
	Overall	2.0171	Constant	2.5671	Constant
Wheat	VI	1.4920	Constant	0.3059	Constant
	VII	3.7999	Constant	0.3841	Constant
	VIII	0.1781	Constant	1.2539	Constant
	Overall	-1.1279	Constant	1.4116	Constant
Sugarcane (suru)	VI	0.9556	Constant	-0.2240	Constant
	VII	1.2302	Constant	-0.1460	Constant
	Overall	1.0574	Constant	0.3772	Constant
Sugarcane (ratoon)	VI	1.0912	Constant	0.7349	Constant
	VII	1.0911	Constant	0.3074	Constant
	Overall	1.1582	Constant	0.4642	Constant

(**Note** All Z* values fall outside the critical region at 5 per cent level of significance, i.e. $-1.96 < Z^* < +1.96$)

The critical examination of the sums of production elasticites of the individual crop revealed that at the overall level, as well as among the different agro-climatic zones, the sum of bi's do not statistically differ from unity as indicated by Z test. It indicated that there existed a constant returns to scale in the production of all the crops under study, during both the time periods. The finding confirms the results of some investigations. Pawar (1978) indicated the constant returns to scale in the crop production on the farms of irrigated and rainfed region in Maharashtra. Kapase (1992) also confirmed that the returns to scale were constant in the production of the individual crops in different agro-climatic zones of Maharashtra.

5.4 Changes in Resource Use efficiency for Different Crops

The estimated production functions underlying the crop production activity enable us to proceed further in an evaluation of the efficiency of prevalent factors of productions in agriculture in different agro-climatic zones. The production function analysis has been generally used to determine economic efficiency of resource use, which requires estimation of marginal value products (MVP) of resources. A resource or input is considered to be used efficiently, if the marginal value product is just sufficient to offset the cost. Equality of marginal value product to factor cost is, therefore, the basis of condition that must be satisfied to obtain efficient resource use.

Accordingly, Cobb-Douglas type of production function has been estimated to examine the resource use efficiency for the crops under study. The variables consider for the production function analysis are:

Y = Output of crop (Qtls.)
X_1 = Human labour (Mandays)
X_2 = Bullock power (Pairdays)
X_3 = Seed (Kg.)
X_4 = Manure and fertilizers (Rs).
X_5 = Other working capital (Rs).

For *ratoon* sugarcane, only four variables viz; human labour, bullock power, manure and fertilizers, and other working capital were considered. The estimates of the production functions for different crops according to agro-climatic zones are given in Appendix IX (A) to IX (I). The geometric means of output and input variables of individual crops across the agro-climatic zones, for both pre-technological and post technological periods i.e. 1980-81 and 2000-01 are given in Appendix- X.

The marginal value products (at geometric mean level) of the individual resources are worked out for selected crops separately for the years 1980-81 and 2000-01 and same are presented in Table 5.11 to 5.19, along with the per unit acquisition costs of the respective resources. The geometric means are commonly used for agricultural data because the distribution of inputs and output is usually positively skewed. Hence, geometric mean being closer to the modes is a more appropriate measure of the central tendency than the arithmetic mean. The difference between marginal value products of the resources and their per unit acquisition costs are tested for their significance with the help of 't' test.

Based on the data given in Tables 6.11 through 6.19, a comparison is made between the marginal value products (at the geometric mean level) of all significant inputs in the equations, with the per unit prices (acquisition costs) to study the changes in farmers allocation efficiency of agricultural resources for different crops over a period under reference.

***Kharif* sorghum :** The marginal value product of human labour was observed non-significant to its price, during both the years i.e. 1980-81 and 2000-01, indicating thereby rational use of human labour for the crop. In other words, there was efficient use of human labour in the production

Table 5.11 : Comparison of Marginal Value Products of Resource Inputs with their Prices in *Kharif* Sorghum

(Rs. per unit of input)

Particulars	1980-81			2000-01		
	VII	VIII	Overall	VII	VIII	Overall
Human labour						
MVP	60.19	47.80	34.77	29.55	35.45	37.72
Price	40.84	43.67	42.40	40.84	43.67	42.40
Bullock power						
MVP	14.03***	99.96***	142.08***	60.05***	87.74***	267.42***
Price	169.19	162.79	166.01	169.19	162.79	166.01
Seed						
MVP	67.67	-76.21***	-64.64***	119.60***	-175.38***	-157.12***
Price	58.81	63.02	61.36	58.81	63.02	61.36
Manures and fertilizers						
MVP	1.40	1.71	1.50	0.99	-0.39	0.48
Price	1.00	1.00	1.00	1.00	1.00	1.00
Other working capital						
MVP	1.00	-14.14	1.72	0.61	1.38	0.37
Price	1.00	1.00	1.00	1.00	1.00	1.00

(Note :- *, ** and *** indicates the level of significance at 10, 5 and 1 per cent, respectively.)

of *kharif* sorghum. The marginal value product of bullock power was found significantly lower than its price during the year 1980-81, whereas, it was significantly higher than its wages during 2000-01. This showed that the use of bullock power during 1980-81 was in excess, while it was less than optimum in 2000-01. The marginal value product of seed was significantly lower than its prices, in both the years, which indicated that this input was used uneconomically in both the years. The difference between marginal value product of manure and fertilizers, and other working capital and their prices were found non-significant, thus, showing the rational use of these inputs during both the years.

Among the agro-climatic zones, the use of bullock power in all the zones and seed in zone VIII during 1980-81 was found uneconomical in the production of *kharif* sorghum, as the marginal value product of these inputs were significantly lower than the unit costs in the respective zones. In the year 2000-01, the use of bullock power was in excess and the use of seed was less than optimum in zone VII and uneconomical in zone VIII. All remaining other resources viz; human labour, manure and fertilizers, and other working capital were used efficiently in respective zones during the period, as the marginal value products of these resources were not significantly differ with their marginal costs.

***Kharif* paddy :** At the overall level, human labour was used efficiently during the years 1980-81, as the marginal value product of this input was non-significant with their marginal cost. The same was used sub-optimally in the year 2000-01. The marginal value product of bullock power was significantly less than its wage rate, indicating the uneconomical use of this resource input during the year 1980-81, while it was significantly higher than the wage rates during the year 2000-01, indicating thereby intensive use of this input. The marginal value product of seed was significantly higher than its unit price, indicating the inefficient use of the seed, during both the periods. The inputs viz; manure and fertilizers, and other working capital were efficiently used during the period, as the difference between the marginal value product and the marginal costs were not significant.

Among the zones, the marginal value products of human labour in all the zones during the year 1980-81 was non-significant with its wage rates, indicating efficient use, whereas, the same was found significantly higher than the wage rates in all the zones during the year 2000-01, indicating inefficient use, that suggested to increase the use level of this resource input. The marginal value product of bullock power in zones II and IX during the year 1980-81 and that for zones V and IX during the year 2000-

Table 5.12 : Comparison of Marginal Value Products of Resource Inputs with their Prices in *Kharif* Paddy

(Rs. per unit of input)

Particulars	*1980-81*				*2000-01*			
	II	*V*	*IX*	*Overall*	*II*	*V*	*IX*	*Overall*
Human labour								
MVP	-14.08	-6.97	19.36	24.59	70.49***	83.18*	92.93**	57.83*
Price	44.47	48.02	36.98	45.30	44.47	48.02	36.98	45.30
Bullock power								
MVP	313.20***	43.53***	290.95***	111.85***	-512.20***	630.87***	1287.09***	202.13***
Price	185.54	171.69	183.05	182.18	185.54	171.69	183.05	182.18
Seed								
MVP	157.44***	218.77***	181.43***	49.35***	-305.73***	138.26**	219.00***	78.31*
Price	17.42	14.24	12.07	15.86	17.42	14.24	12.07	15.86
Manures and fertilizers								
MVP	6.67	1.72	0.61	3.98	-0.73	2.66	3.88	2.69
Price	1.00	1.00	1.00	1.00	1.00	1.00	1.00	1.00
Other working capital								
MVP	-0.99	0.48	25.40***	3.12	4.34	10.48	8.55	1.54
Price	1.00	1.00	1.00	1.00	1.00	1.00	1.00	1.00

(Note :- *, ** and *** indicates the level of significance at 10, 5 and 1 per cent, respectively.)

01 was significantly higher than the wage rates of bullock pair. The use of seed was sub-optimal in all the zones during 1980-81, whereas, the same was used sub-optimally in zones V and IX, during the later period. The manure and fertilizers was efficiently used in all the zones, during the period, as the difference between marginal value product and marginal cost of this resource input was not significant. The marginal value product of other working capital in zone IX, during the year 1980-81 was significantly higher than its marginal cost. There existed scope to use more capital.

Pearlmillet : The marginal value products of human labour, manure and fertilizers and other working capital were non-significant during both the years, indicating the rational use of these resources, at the overall level. The marginal value product of bullock power was significantly lower than its unit price during base period, while the reverse was the true during the later period. The marginal value product of seed was significantly higher than its unit price, during the base year i.e. 1980-81, whereas, it was turn out to be significantly negative during the later period.

The agro-climatic zonewise comparison of the estimates of the different resources with their unit prices indicated uneconomical use of bullock power and seed in zones V and VII, and manure and fertilizers in zone VI, were noticed. While, sub-optimal use of bullock power and seed was found in zone VI, and manure and fertilizers in zone V, during the year 1980-81. Similarly, during the later period, the use of bullock power and seed in all the zones was uneconomical as the marginal value products of these resources were significantly lower than their unit prices. The use of human labour in zone VI was found to be less than optimum. All other remaining resources were used efficiently in the respective zone during the later period.

Groundnut : At the overall level, the use of human labour was rational in 1980-81, however, in the year 2000-01, its use was found less than optimal in the production of groundnut crop. The use of bullock power was excess during 1980-81, as its marginal value product was significantly lower than the bullock-pair wage rates, whereas, it was significantly higher with its per unit charges during the later period, indicating the intensive use of bullock power. This showed that there existed a scope for reallocation of the funds in the favour of this input factor during the period. The use of seed was found excess in the year 1980-81, whereas, it was used rationally during the later period.

Among the agro-climatic zones, the resource use was found to be less than optimum in production of groundnut crop in the case of bullock

Table 5.13 : Comparison of Marginal Value Products of Resource Inputs with their Prices in Pearlmillet

(Rs. per unit of input)

Particulars	*1980-81*				*2000-01*			
	V	*VI*	*VII*	*Overall*	*V*	*VI*	*VII*	*Overall*
Human labour								
MVP	33.15	43.58	46.58***	40.07	35.19	65.58***	41.87	47.92
Price	47.40	45.74	42.91	45.55	47.40	45.74	42.91	45.55
Bullock power								
MVP	17.64***	321.73***	35.02***	85.11***	51.95***	41.80***	-132.85***	347.02**
Price	175.00	157.75	165.29	165.36	175.00	157.75	165.29	165.36
Seed								
MVP	-15.88***	210.02***	7.48***	225.59***	-947.03***	-449.47***	39.28***	-211.74***
Price	85.36	67.97	72.80	73.94	85.36	67.97	72.80	73.94
Manures and fertilizers								
MVP	6.47***	-10.00***	0.16	-1.41	3.07	-0.28	0.37	0.27
Price	1.00	1.00	1.00	1.00	1.00	1.00	1.00	1.00
Other working capital								
MVP	-0.90	2.14	-0.41	-0.91	-0.32	1.37	3.26*	1.73
Price	1.00	1.00	1.00	1.00	1.00	1.00	1.00	1.00

(Note :- *, ** and *** indicates thc level of significance at 10, 5 and 1 per cent, respectively.)

Table 5.14 : Comparison of Marginal Value Products of Resource Inputs with their Prices in Groundnut

(Rs. per unit of input)

Particulars	*1980-81*			*2000-01*		
	V	*VII*	*Overall*	*V*	*VII*	*Overall*
Human labour						
MVP	143.47	110.23	61.87	93.69	102.36***	87.54***
Price	51.62	45.05	48.38	51.62	45.05	48.38
Bullock power						
MVP	1201.73***	-349.54***	-230.74***	-288.63***	1583.81***	1050.42***
Price	201.62	160.67	178.58	201.62	160.67	178.58
Seed						
MVP	320.00*	-48.54	-14.80*	-99.37***	-149.64	17.42
Price	33.11	28.47	30.33	33.11	28.47	30.33
Manures and fertilizers						
MVP	1.20	2.01	1.96	1.03	1.17	2.02
Price	1.00	1.00	1.00	1.00	1.00	1.00
Other working capital						
MVP	-3.27	4.51*	3.62	1.15	1.16	0.42
Price	1.00	1.00	1.00	1.00	1.00	1.00

(Note :- *, ** and *** indicates the level of significance at 10, 5 and 1 per cent, respectively.)

power and seed in zone V and other working capital in zone VII during the year 1980-81. The resources such as human labour and bullock power were observed to be less than optimum in zone VII, whereas, the excess use of bullock power and seed in zone V was observed, during the later period. The manure and fertilizers, and other working capital were observed to be rationally used in all the zones, during both the time periods, excepting other working capital in zone VII during the year 1980-81.

Cotton : In the production of cotton crop, at the overall level, the use of human labour was observed to be inefficient during the period. However, the decrease in marginal value productivity of human labour (from Rs. 101.02 to Rs. 76.06) roughly indicated that the productivity of this resource has decreased considerably during the period. The use of bullock power and seed in the year 1980-81 was found excess, whereas, in 2000-01 the use of these inputs was less than the optimum. The remaining resource inputs were found efficiently used during the period.

Between the two agro-climatic zones under study, the resource use was sub-optimal in the case of human labour in zone VII, during both the time periods. The use of bullock power was inefficient only in zone VIII, in the year 2000-01, while its use was excess in other zones. Similarly, the marginal value of seed was significantly less than its unit price in both the zones during the period, except in zone VII in the year 2000-01, indicating uneconomical use of this input. The marginal value products of manure and fertilizer, and other working capital were non-significant in all the zones, which indicated the rational use of these resources in the production of cotton crop.

***Rabi* sorghum :** At the overall, the use of human labour, bullock power and seed were found to be excess during both the periods, whereas, that for manure and fertilizers, the use of this input was less than optimum. The use of other working capital was rational.

Amongst the two agro-climatic zones, the use of human labour was uneconomical in zone VI, whereas, it was rational in zone VII during both the years. The use of bullock power in both the zones, i.e. in zones VI and VII in the year 1980-81 was less than optimum, whereas, it was used uneconomically in the said zones during the later period. The use of seed input during both the periods was in excess, except in zone VI during the year 2000-01. The use of manure and fertilizers was found rational in zone VI during both the periods, whereas, in zone VII, it was less than optimum in the year 1980-81 and uneconomical in the year 2000-01. The use of other working capital was rational in all the zones, during the period. It indicated that there was a scope to reallocate funds for the use of those resources,

Table 5.15 : Comparison of Marginal Value Products of Resource Inputs with their Prices in Cotton

(Rs. per unit of input)

Particulars	*1980-81*			*2000-01*		
	VII	*VIII*	*Overall*	*VII*	*VIII*	*Overall*
Human labour						
MVP	127.09***	72.22	101.02***	22.92	10.67*	76.06*
Price	40.36	41.09	40.77	40.36	41.09	40.77
Bullock power						
MVP	-7.12***	171.14**	125.89***	-62.32***	632.56***	201.93***
Price	168.56	187.72	180.52	168.56	187.72	180.52
Seed						
MVP	-183.13***	-602.19***	-355.35***	1221.77***	-271.80***	298.67***
Price	277.69	177.53	196.68	277.69	177.53	196.68
Manures and fertilizers						
MVP	2.80	1.83	1.71	1.03	0.46	0.34
Price	1.00	1.00	1.00	1.00	1.00	1.00
Other working capital						
MVP	-3.78	3.16	1.32	-1.02	2.69	1.35
Price	1.00	1.00	1.00	1.00	1.00	1.00

(Note :- *, ** and *** indicates the level of significance at 10, 5 and 1 per cent, respectively.)

Table 5.16 : Comparison of Marginal Value Products of Resource Inputs with their Prices in *Rabi* Sorghum

(Rs. per unit of input)

Particulars	*1980-81*			*2000-01*		
	VI	*VII*	*Overall*	*VI*	*VII*	*Overall*
Human labour						
MVP	23.17***	42.84	24.16***	15.36**	31.59	14.39***
Price	47.93	44.99	46.25	47.93	44.99	46.25
Bullock power						
MVP	906.28***	281.81***	-128.41***	-0.92***	-47.18***	-192.16***
Price	162.55	130.27	141.27	162.55	130.27	141.27
Seed						
MVP	-287.96***	-311.22***	-530.41***	1034.67***	11.65	-586.42***
Price	9.92	19.74	15.41	9.92	19.74	15.41
Manures and fertilizers						
MVP	0.29	14.39***	15.54***	4.19	-24.38***	27.00**
Price	1.00	1.00	1.00	1.00	1.00	1.00
Other working capital						
MVP	-1.80	-1.43	2.52	0.88	2.94	-0.53
Price	1.00	1.00	1.00	1.00	1.00	1.00

(Note :- *, ** and * indicates the level of significance at 10, 5 and 1 per cent, respectively.)**

Table 5.17 : Comparison of Marginal Value Products of Resource Inputs with their Prices in Wheat

(Rs. per unit of input)

Particulars	*1980-81*				*2000-01*			
	VI	*VII*	*VIII*	*Overall*	*VI*	*VII*	*VIII*	*Overall*
Human labour								
MVP	156.79***	116.66***	7.52	77.71	27.54	26.14	37.97	59.48
Price	47.78	46.34	49.58	48.06	47.78	46.34	49.58	48.06
Bullock power								
MVP	-139.43***	138.69	-76.55***	-304.67***	88.68***	174.83***	213.33***	22.86***
Price	154.00	153.76	119.82	130.74	154.00	153.76	119.82	130.74
Seed								
MVP	129.70	537.15*	75.75	-357.84*	23.67	-43.35	24.55	14.41
Price	11.75	11.19	11.97	11.71	11.75	11.19	11.97	11.71
Manures and fertilizers								
MVP	-0.24	-0.23	0.35	1.37	1.16	2.70	1.07	4.51*
Price	1.00	1.00	1.00	1.00	1.00	1.00	1.00	1.00
Other working capital								
MVP	0.83	0.02	0.02	0.84	2.60	1.90	2.08	1.52
Price	1.00	1.00	1.00	1.00	1.00	1.00	1.00	1.00

(Note :- *, ** and * indicates the level of significance at 10, 5 and 1 per cent, respectively.)**

which were sub-optimally used in order to have higher production of *rabi* sorghum.

Wheat : At the overall level, the resource use of human labour was rational during both the time periods, whereas, that of for bullock power, its use was in excess. In the case of seed, the excess use was found in the year 1980-81 and it was rationally used during the year 2000-01. The use of manure and fertilizers was rational in the base period, while it was used sub-optimally during the later period. The use of other working capital was rational during both the periods.

Among the agro-climatic zones, the use of human labour in zones VI and VII was less than the optimum in the year 1980-81, whereas, it was found rational in all the zones in the year 2000-01. In the case of bullock power, the use was excess in zones VI and VIII, during the year 1980-81 and in zone VI during the year 2000-01, whereas, it was used sub-optimally in zones VII and VIII during later period. The marginal value product of seed was significantly higher than its marginal costs in zone VII during the year 1980-81, while it was found non-significant in all the zones during the year 2000-01. The efficient use of manure and fertilizers, and other working capital was found in all the zones during both the periods.

***Suru* Sugarcane :** Amongst all the resource inputs, the use of bullock power was in excess during the year 1980-81, while, it was used sub-optimally during the later period. The marginal value products of all the remaining variables were not significantly differed from their marginal costs during both the periods, at the overall level. This indicated that all the resources were used rationally during the period under study.

Between the two agro-climatic zones, the marginal value product of human labour was found significantly lower in both the zones during the year 2000-01, indicating the excess use of this input. The bullock power was used uneconomically in all the zones, as the marginal value product of this input was significantly lower than its wage rate. All other resource inputs were not significantly differed from their marginal costs, indicating that these resources were used rationally in all the zones, during the period under study. There existed a limited scope to increase production of *suru* sugarcane through reallocation of existing resources on the basis of magnitudes of their marginal value products.

***Ratoon* Sugarcane :** The farms were found to be most efficient in the production of *ratoon* sugarcane, as the marginal value product of all the resource inputs (excepting human labour and bullock power during the year 2000-01) were not significantly differed from their marginal costs, at the overall level.

Table 5.18 : Comparison of Marginal Value Products of Resource Inputs with their Prices in *Suru* Sugarcane

(Rs. per unit of input)

Particulars	*1980-81*			*2000-01*		
	VI	*VII*	*Overall*	*VI*	*VII*	*Overall*
Human labour						
MVP	58.66	11.37	36.76	-71.74***	-16.46*	75.79
Price	51.32	53.78	52.52	51.32	53.78	52.52
Bullock power						
MVP	-4.66***	72.70***	123.89***	162.55***	97.39***	784.83***
Price	188.72	167.18	178.69	188.72	167.18	178.69
Seed						
MVP	0.17	1.55	0.66	-3.19	-3.40	1.07
Price	1.04	1.05	1.04	1.04	1.05	1.04
Manures and fertilizers						
MVP	1.33	0.29	0.27	0.80	-0.26	0.76
Price	1.00	1.00	1.00	1.00	1.00	1.00
Other working capital						
MVP	0.10	-0.11	0.48	-0.13	-0.25	1.04
Price	1.00	1.00	1.00	1.00	1.00	1.00

(Note :- *, ** and *** indicates the level of significance at 10, 5 and 1 per cent, respectively.)

Table 5.19 : Comparison of Marginal Value Products of Resource Inputs with their Prices in *Ratoon* Sugarcane

(Rs. per unit of input)

Particulars	*1980-81*			*2000-01*		
	VI	*VII*	*Overall*	*VI*	*VII*	*Overall*
Human labour						
MVP	43.85	20.53	50.74	76.26	100.98**	82.95**
Price	48.96	51.12	50.37	48.96	51.12	50.37
Bullock power						
MVP	3.55***	313.58***	72.20***	199.26***	117.20***	294.98***
Price	146.62	184.80	177.33	146.62	184.80	177.33
Manures and fertilizers						
MVP	0.32	0.17	0.50	0.58	0.31	0.80
Price	1.00	1.00	1.00	1.00	1.00	1.00
Other working capital						
MVP	0.25	-0.59	-0.04	1.40	-0.51	-0.37
Price	1.00	1.00	1.00	1.00	1.00	1.00

(Note :- *, ** and *** indicates the level of significance at 10, 5 and 1 per cent, respectively.)

Across the agro-climatic zones, it is seen that the use of human labour was less than optimum in zone VII in the year 2000-01. The use of bullock power was found excess in zone VI and less than optimum in zone VII during the year 1980-81, whereas, it was used sub-optimally in zone VI and excess in production of major crops across the agro-climatic zones indicated that the marginal value products of inputs were significantly different from their unit prices in 19 cases out of 44 comparisons made, at the overall level and 43 cases out of 103 cases in the different agro-climatic zones during the year 1980-81. While, the same were differed from their unit prices in 22 cases out of 44 comparisons made, at the overall level and 45 cases out of 103 cases in the different agro-climatic zones during the year 2000-01. These findings suggested that the degree of efficiency of allocation of agricultural resources was relatively higher in 2000-01 than that in 1980-81. This is because of adoption of crop production and management technology during the later period. This particular finding clearly indicated that, there was great disequilibrium during the period under study, in resource use efficiency in all the crops and across the agro-climatic zones that too.

Since, the difference between the marginal value products of manure and fertilizers, and other working capital and their marginal costs were non-significant in majority of crops during both the years. It was clear that the sample farms were efficiently using these resource inputs at optimal level in both the time periods.

According to Cavery (1991), the productivity of groundnut in Tamil Nadu was increased during the period of 22 years (i.e. 1965-66 to 1987-88), due to changes in resources use pattern as consequences of new Technology. Kapase (1992) concluded that, the per hectare resource use productivity of plant nutrients, human labour and quality of seeds have influenced the output of major crops and the agro-climatic differences led wide variations in per hectare resource use of individual crops. Raut (1997) also confirmed that the per hectare resource use levels for most of the crops were higher and there was substantial increase in per hectare input utilization during the period from 1973-74 to 1991-92. Reddy (2002) showed that in paddy cultivation in Andhra Pradesh, the use of labour and seed declined during the period of 1981-82 to 1991-92. Thus, the findings of the resource productivities and resources use efficiency in production of major crops are consistent with the study results of Cavery (1991), Kapase (1992), Raut (1997) and Reddy (2002). Hence, the hypothesis viz. there existed changes in resource productivities and resource use efficiencies in the production of major crops over the period, has been proved.

5.5 Economics of Optimal Use Levels of Resource Variables

It has been observed in the previous section that there existed some inefficiency in the use of resource inputs in the production of crops in different agro-climatic zones in both the years. In recognition of the dis-equilibrium in resource use efficiency, an attempt was made to determine economic optimal use levels of the various resource inputs in the production of crops, under limited capital (actual used capital) situation, at two points of times and the changes therein over a period of time were studied. The very purpose of such analysis is to examine the extent to which existing level of input use deviated from the optimal level and thereby to find out possible increase in the output level of individual crops through reallocation of the available capital. The analysis regarding estimation of optimal level use of farm resources for different crops has also supported the fact that there prevailed the possibilities of increasing output and gross returns of individual crops in both the years through optimal reallocation of farm resources.

The economic optimal levels of farm inputs viz; human labour, bullock power, seed, manure and fertilizers, and other working capital have been estimated separately for individual crops across the agro-climatic zones in both the years on per hectare basis. The constraint of limited capital has been imposed with a view that the estimates merely indicate the potential of increasing crop output, but the availability of the required capital does not permit immediate drastic change in the resource use pattern in the direction of optimization of farm resources. The estimated existing and optimal levels of farm inputs and output per hectare under limited capital situation are given in Table 5.20 to Table 5.28, separately for individual crops and across the different agro-climatic zones that too, for the pre-technological and post technological period, i.e. 1980-81 and 2000-01.

***Kharif* sorghum :** The estimates relating to optimal use levels of resources in the production of *kharif* sorghum at the overall level, indicated that the funds should have been reallocated more human labour, bullock power and other working capital by reducing manure and fertilizers, during the year 1980-81, to obtain 17.04 per cent higher output, whereas, reduction in allocation of fund on human labour, manure and fertilizers, and other working capital during the year 2000-01, could obtain 13.69 per cent higher output over the existing level.

During the year 1980-81, the difference between optimal and existing levels of inputs obtained for zone VII indicated the possibility of increasing the existing level of output by 20.98 per cent through reallocation of funds in favour of more human labour and seed. As against this, there

Table 5.20 : Comparison of Existing and Optimal Use Levels of Resource Inputs and Output of *Kharif* Sorghum

Particulars	*1980-81*			*2000-01*		
	VII	*VIII*	*Overall*	*VII*	*VIII*	*Overall*
Capital allocation (Rs.)	7679.94	4813.60	5602.77	6559.67	6154.58	6337.39
Human labour (mandays)						
Existing level	93.77	50.31	62.26	87.83	84.10	85.83
Optimal level	227.48	116.06	89.51	84.79	162.10	75.63
Bullock power (pairdays)						
Existing level	18.35	12.43	14.05	14.30	10.92	12.34
Optimal level	10.52	15.84	19.71	5.48	8.43	13.64
Seed (kgs.)						
Existing level	11.47	--	--	8.37	--	--
Optimal level	30.52	--	--	25.36	--	--
Manures and fertilizers (Rs.)						
Existing level	2993.60	295.43	1026.45	2279.25	--	1710.61
Optimal level	2440.23	187.68	473.23	2879.71	--	989.27
Other working capital (Rs.)						
Existing level	387.98	--	227.25	1044.30	1450.01	1273.07
Optimal level	386.49	--	323.47	1083.00	4693.33	436.76
Output (Qtl.)						
Existing level	15.30	7.11	9.33	22.95	16.95	19.57
Optimal level	18.51	8.18	10.92	27.24	17.30	22.25
Percentage increase	20.98	15.05	17.04	18.69	2.06	13.69

existed the possibility to obtain 15.05 per cent higher output over the existing one through reallocation of funds more human labour and bullock power resources by reducing on manure and fertilizers in zone VIII. In the later period, the difference between optimal and existing use levels of inputs obtained in zone VII indicated that the funds should have been reduced on the resources viz., human labour and bullock power inputs and reallocate more on seed, manure and fertilizers and other working capital to obtain 18.69 per cent higher output over the existing level of output. The estimates relating to the optimal resource use levels in the production of *kharif* sorghum in zone VIII, during the year 2000-01 indicated that a small increase in the output (2.06 %) was possible by reducing funds on bullock power and reallocate more on human labour and other working capital.

***Kharif* paddy :** The estimated optimal use levels of resource in the production of *kharif* paddy at the overall level, indicated the need to divert available funds from human labour and bullock power, and reallocate more in favour of seed, manure and fertilizers, and other working capital to obtain an increase in the existing level of output by 11.74 per cent, during the year 1980-81. However, the difference between optimal and existing uses of all the inputs during the later indicated that there was no scope for reallocation of funds. By reducing the expenditure on these resources only 3.40 per cent increase in the existing level of output is possible.

Among the zones, during the year 1980-81, the estimates relating to optimal resource use for zone II indicated the need to divert the funds in favour of more bullock power, seed and manure and fertilizers to obtain 4.45 per cent higher output. The zone V indicated the need to reallocate the funds in favour of more bullock power, seed, manure and fertilizers and other working capital to obtain 18.06 per cent higher output. The zone IX indicated a need to divert the funds from human labour and manure and fertilizers and reallocate more on bullock power, seed and other working capital to obtain 9.16 per cent higher output than the existing level. In the year 2000-01, there was a need for reallocation of the funds in favour of more human labour and other working capital in zones II. The zone V showed that the funds should be reduced from bullock power, seed and manure and fertilizers, and reallocated in favour of more human labour and other working capital to obtain small increase of 1.21 per cent output. Similarly, in zone IX, there was less scope to reallocate the available funds, as the differences between optimal and existing use of inputs were less than unity. With the small adjustment of funds in favour of more other working capital could increase only 1.11 per cent output.

Table 5.21 : Comparison of Existing and Optimal use Levels of Resource Inputs and Output of *Kharif* Paddy

Particulars	*1980-81*				*2000-01*			
	II	*V*	*IX*	*Overall*	*II*	*V*	*IX*	*Overall*
Capital allocation (Rs.)	7960.75	11941.79	6499.37	8089.00	4944.63	5753.31	8609.31	5632.50
Human labour (mandays)								
Existing level	- -	- -	111.26	119.03	40.90	61.82	138.58	55.91
Optimal level	- -	- -	49.47	69.99	183.36	74.78	135.91	40.90
Bullock power (pairdays)								
Existing level	11.66	11.81	10.04	10.81	- -	7.99	11.83	8.72
Optimal level	26.34	30.74	14.08	6.72	- -	4.84	8.86	2.09
Seed (kgs.)								
Existing level	44.32	46.68	44.35	44.55	- -	93.52	101.84	91.26
Optimal level	168.74	118.15	112.64	132.16	- -	61.62	50.95	65.84
Manures and fertilizers (Rs.)								
Existing level	529.71	2476.03	523.05	770.40	- -	792.82	1499.78	1976.93
Optimal level	3603.52	7581.95	135.13	2300.27	- -	757.09	994.31	1717.65
Other working capital (Rs.)								
Existing level	- -	794.34	152.45	721.59	825.28	1700.17	767.95	1017.40
Optimal level	- -	1142.22	1261.99	773.31	1973.56	2108.38	2529.99	348.29
Output (Qtl.)								
Existing level	9.44	10.80	13.54	11.75	20.34	29.83	23.35	22.95
Optimal level	9.86	12.75	14.78	13.13	20.59	30.19	23.61	23.73
Percentage increase	4.45	18.06	9.16	11.74	1.23	1.21	1.11	3.40

Table 5.22 : Comparison of Existing and Optimal Use Levels of Resource Inputs and Output of Pearlmillet

Particulars	*1980-81*				*2000-01*			
	V	*VI*	*VII*	*Overall*	*V*	*VI*	*VII*	*Overall*
Capital allocation (Rs.)	3850.85	4216.11	4447.01	4086.57	6486.19	4020.97	4100.98	4680.56
Human labour (mandays)								
Existing level	28.26	43.12	45.97	37.54	69.55	44.94	57.04	54.38
Optimal level	64.94	67.57	54.87	45.14	53.74	127.76	96.04	54.12
Bullock power (pairdays)								
Existing level	12.60	12.69	13.37	12.77	16.06	10.74	- -	11.43
Optimal level	7.70	6.31	18.23	8.56	4.61	5.71	- -	19.04
Seed (kgs.)								
Existing level	- -	3.51	3.57	3.54	- -	- -	3.97	- -
Optimal level	- -	4.71	3.98	4.90	- -	- -	3.56	- -
Manures and fertilizers (Rs.)								
Existing level	5.12	- -	218.04	- -	695.33	- -	593.90	613.48
Optimal level	33.15	- -	21.77	- -	499.02	- -	124.01	34.60
Other working capital (Rs.)								
Existing level	- -	165.06	- -	- -	- -	452.48	399.21	471.44
Optimal level	- -	318.46	- -	- -	- -	1314.86	2400.29	765.47
Output (Qtl.)								
Existing level	2.60	4.21	6.44	3.93	16.00	9.46	9.11	11.07
Optimal level	2.75	4.88	6.56	4.04	16.34	11.44	9.54	12.08
Percentage increase	5.77	15.91	1.86	2.80	2.13	20.93	4.72	9.12

Pearlmillet : The estimated optimal levels of resources in the production of pearlmillet, at the overall level, indicated the need for higher level use of human labour and seed in 1980-81 and that for bullock power and other working capital in 2000-01; and less use of bullock power in 1980-81 and that of human labour and manure and fertilizers in 2000-01 for obtaining 2.80 and 9.12 per cent increase in existing level output, at the overall level during the respective years.

The comparison of the optimal and existing use levels of inputs in zone V, indicated the need of less expenditure on bullock power and reallocate the more funds in favour of human labour and manure and fertilizers in 1980-81; and less on human labour, bullock power and manure and fertilizers in 2000-01. This could have contributed to increase in the output by 5.77 and 2.13 per cent over the existing output level during the respective years. In zone VI, by reallocating funds in favour of more human labour, seed and other working capital and by reducing funds on bullock power in 1980-81 and; reallocating more funds on human labour and other working capital by reducing funds of bullock power in 2000-01, could have contributed to 15.91 and 20.93 per cent increase in the existing level of output in respective time periods. Similarly, in the case of zone VII, it is found that there was the possibility of obtaining 1.86 per cent increase in the existing output level, by reallocating the available funds more in favour of human labour, bullock power and seed, and by reducing funds from manure and fertilizers in 1980-81. In the later period, by minimizing expenditure on seed, and manure and fertilizers and reallocating more in favour of human labour and other working capital in 2000-01, could increase the existing level of output by 4.72 per cent in zone VII.

Groundnut : At the overall level, the comparison of optimal output with existing levels indicated that the available funds should have been used in favour of more human labour, manure and fertilizers and other working capital in 1980-81 and that for less bullock power, seed and other working capital, and more human labour and manure and fertilizers in 2000-01, in order to maximize output. These adjustments could have resulted in increasing the output by 8.98 and 7.79 per cent in the year 1980-81 and 2000-01, respectively.

Among the zones, during the base period, the difference between optimum and existing use levels of inputs suggested that the funds should have been reallocated in favour of more human labour, bullock power, seed and manure and fertilizers in zone V and; reallocating more to human labour, manure and fertilizers and other working capital in zone VII, to obtain 6.41 and 2.83 per cent increase in the existing level of output in the respective zones. As against this, during the later period, there existed the possibility of obtaining 9.69 and 5.38 per cent increase in the existing level of output in these two zones, by diverting funds more human labour, manure and fertilizers and other working capital in zone V and; by reallocating available funds in favour of more human labour, bullock power and, manure and fertilizers, and other working capital in zone VII.

Cotton : The estimated optimal levels of resources in the production of cotton crop, at the overall level, it was noticed that there was very less scope to reallocate the funds, as the optimal output could be increased by only 0.57 per cent of existing level in 1980-81, which could be done by higher use of human labour and less use of bullock power, manure and fertilizers and other working capital. In the later period, an increase of 1.60 per cent in existing output was possible by reallocating available funds in favour of more use of human labour and seed, and less use of bullock power, manure and fertilizers and other working capital than their existing levels of use to optimize the output.

Though, the comparison of the optimal and existing use levels of inputs in the production of cotton crop during the year 1980-81 revealed that a marginal increase in output was possible at the overall, the individual zones possessed a scope to increase output of cotton through reallocation of resources. The difference between optimum and existing use levels of inputs obtained for zone VII suggested the need for reallocation of funds in favour of more human labour and less manure and fertilizers in 1980-81, whereas, the reallocation of funds should have been in favour of more human labour, seed and manure and fertilizers in this zone in 2000-01, to obtain 2.33 and 6.22 per cent increase in the existing levels of output in respective years. The comparison of optimal output with the existing levels in zone VIII, indicated that the available funds should have been reallocated in favour of more human labour, bullock power and other working capital and less manure and fertilizers, in 1980-81 and; more bullock power and other working capital and less human labour and manures and fertilizers, in 2000-01, to obtain 4.96 and 5.11 per cent higher output of cotton during respective years.

***Rabi* sorghum :** The estimates relating to optimal resource use levels in the production of *rabi* sorghum, at the overall level, indicated that increase in output (8.50 %) was possible in 1980-81 through reallocation of available capital in favour of more manure and fertilizers and other working capital and less human labour. The difference between optimal and existing levels of inputs obtained for the year 2000-01 indicated that there was little scope for reallocation of existing resources. However, the funds should

Table 5.23 : Comparison of Existing and Optimal Use Levels of Recourse Inputs and Output of Groundnut

Particulars	*1980-81*			*2000-01*		
	V	*VII*	*Overall*	*V*	*VII*	*Overall*
Capital allocation (Rs.)	12699.51	9704.76	10971.29	13372.48	8697.77	10566.96
Human labour (mandays)						
Existing level	100.77	89.11	94.09	121.76	76.48	94.23
Optimal level	147.93	550.33	349.85	581.36	550.81	162.84
Bullock power (pairdays)						
Existing level	18.93	- -	- -	- -	14.47	15.85
Optimal level	22.83	- -	- -	- -	18.00	9.22
Seed (kgs.)						
Existing level	111.80	- -	- -	- -	- -	102.06
Optimal level	138.09	- -	- -	- -	- -	79.82
Manures and fertilizers (Rs.)						
Existing level	2654.36	973.06	1991.68	3903.81	1285.02	2311.96
Optimal level	3132.56	5987.52	6015.01	9350.29	2872.56	2696.43
Other working capital (Rs.)						
Existing level	- -	433.03	864.98	1680.28	911.58	1213.02
Optimal level	- -	9978.29	6319.46	4187.61	2001.47	298.24
Output (Qtl.)						
Existing level	13.26	5.66	8.91	18.68	8.00	12.19
Optimal level	14.11	5.82	9.71	20.49	8.43	13.14
Percentage increase	6.41	2.83	8.98	9.69	5.38	7.79

Table 5.24 : Comparison of Existing and Optimal Use Levels of Resource Inputs and Output of Cotton

Particulars	*1980-81*			*2000-01*		
	VII	*VIII*	*Overall*	*VII*	*VIII*	*Overall*
Capital allocation (Rs.)	5279.28	5043.52	4998.95	7133.08	5982.97	6338.39
Human labour (mandays)						
Existing level	48.29	49.48	48.78	96.69	73.55	82.14
Optimal level	126.45	85.63	103.67	122.84	44.47	126.97
Bullock power (pairdays)						
Existing level	- -	13.05	11.83	- -	12.30	12.39
Optimal level	- -	18.98	10.97	- -	22.69	11.98
Seed (kgs.)						
Existing level	- -	- -	- -	3.64	- -	3.30
Optimal level	- -	- -	- -	15.58	- -	4.99
Manures and fertilizers (Rs.)						
Existing level	2727.03	2801.81	2477.80	1951.95	2482.97	2244.78
Optimal level	1156.33	455.22	639.81	2178.25	1917.39	1357.21
Other working capital (Rs.)						
Existing level	- -	468.10	351.76	- -	516.63	648.14
Optimal level	- -	1168.35	236.39	- -	2387.85	539.96
Output (Qtl.)						
Existing level	3.44	3.63	3.52	7.56	5.48	6.25
Optimal level	3.52	3.81	3.54	8.03	5.76	6.35
Percentage increase	2.33	4.96	0.57	6.22	5.11	1.60

Table 5.25 : Comparison of Existing and Optimal Use Levels of Resource Inputs and Output of *Rabi* Sorghum

Particulars	*1980-81*			*2000-01*		
	VI	*VII*	*Overall*	*VI*	*VII*	*Overall*
Capital allocation (Rs.)	2423.68	2567.45	2428.98	4287.62	4823.01	4580.29
Human labour (mandays)						
Existing level	21.89	24.23	22.80	67.03	75.23	71.49
Optimal level	72.68	51.54	20.85	55.59	84.38	102.48
Bullock power (pairdays)						
Existing level	7.78	9.79	--	--	--	--
Optimal level	16.86	19.53	--	--	--	--
Seed (kgs.)						
Existing level	--	--	--	11.07	11.71	--
Optimal level	--	--	--	35.67	17.17	--
Manures and fertilizers (Rs.)						
Existing level	86.76	169.15	118.65	838.44	--	457.45
Optimal level	35.51	1486.87	517.75	1406.94	--	2363.10
Other working capital (Rs.)						
Existing level	--	--	227.49	998.73	444.95	--
Optimal level	--	--	1452.42	1676.04	2336.56	--
Output (Qtl.)						
Existing level	6.05	8.67	7.06	11.60	9.60	10.55
Optimal level	7.55	9.65	7.66	11.66	9.84	10.57
Percentage increase	24.79	11.30	8.50	0.52	2.50	0.19

have been reallocated in favour of more human labour and manure and fertilizers to obtain small increase in output i.e. 0.19 per cent than the existing level of output.

The comparison of the optimal and existing resource use levels in zone VI indicated that the possibility to obtain 24.79 per cent higher output through reallocation of funds in favour of more human labour and bullock power and, less manure and fertilizers, in 1980-81. As against, there existed a possibility of obtaining only 0.52 per cent increase in the existing output level in this zone in 2000-01, through reallocation of funds in favour of more seed, manure and fertilizers, and other working capital and less human labour. In zone VII, the reallocation of funds in favour of more human labour, bullock power, and manure and fertilizers, in 1980-81; and more human labour, seed and other working capital in 2000-01, could have contributed 11.30 and only 2.50 per cent higher output of *rabi* sorghum in respective years.

Wheat : The comparison of the optimal and existing levels of inputs, at the overall level, indicated that more investment in human labour and manure and fertilizers and; less other working capital could have contributed to increase in the existing level output by 5.24 per cent in 1980-81. As against this, there existed the possibility of obtaining 1.92 per cent increase in the existing level of output in 2000-01, through reallocation of funds in favour of more human labour, seed, manure and fertilizers and other working capital and less bullock power.

Among the zones, the differences between the optimal and existing use levels during the year 1980-81, indicated the need for reallocation of funds in favour of more human labour and seed, and less other working capital in zone VI; more human labour, bullock power and seed, and less other working capital in zone VII; and more human labour and seed, and less manure and fertilizers, and other working capital in zone VIII, to obtain 8.54, 7.57 and 11.76 per cent increase in the existing level output, in respective zones. In the year 2000-01, there existed the possibility of obtaining 2.44, 6.64 and 5.14 per cent increase in the existing level of output in respective zones, through reallocation of available funds in favour of more seed and other working capital and, less human labour, bullock power and, manure and fertilizers in zones VI; more in favour of human labour and bullock power, manure and fertilizers and other working capital in zone VII; more in favour of seed and other working capital and less human labour, bullock power and, manure and fertilizers in zone VIII.

***Suru* sugarcane :** In the production of *suru* sugarcane crop, the comparison of optimal and existing levels of output indicated that there

Table 5.26 : Comparison of Existing and Optimal Use Levels of Resource Inputs and Output of Wheat

Particulars	*1980-81*				*2000-01*			
	VI	*VII*	*VIII*	*Overall*	*VI*	*VII*	*VIII*	*Overall*
Capital allocation (Rs.)	6712.33	6860.92	6302.03	6474.94	7173.29	6377.56	6666.77	6695.78
Human labour (mandays)								
Existing level	63.08	66.10	63.45	64.35	97.62	91.69	72.40	83.44
Optimal level	131.35	177.23	179.08	124.51	57.09	170.76	57.43	89.38
Bullock power (pairdays)								
Existing level	- -	16.21	- -	- -	7.69	5.63	14.06	10.28
Optimal level	- -	16.50	- -	- -	3.60	17.06	8.35	2.89
Seed (kgs.)								
Existing level	106.44	106.14	109.14	- -	105.64	- -	105.88	105.20
Optimal level	127.59	108.68	119.41	- -	134.67	- -	118.86	109.70
Manures and fertilizers (Rs.)								
Existing level	- -	- -	990.09	1731.51	1915.53	2976.88	1767.02	2136.87
Optimal level	- -	- -	759.06	2339.20	423.77	4717.38	1668.14	2516.56
Other working capital (Rs.)								
Existing level	4096.19	3464.03	2612.57	3456.94	2123.87	2019.09	1561.00	1814.94
Optimal level	3366.72	74.25	271.17	2419.90	5840.29	12713.40	2930.41	2301.01
Output (Qtl.)								
Existing level	15.10	17.83	8.08	14.32	22.95	19.44	16.54	18.79
Optimal level	16.39	19.18	9.03	15.07	23.51	20.73	17.39	19.15
Percentage increase	8.54	7.57	11.76	5.24	2.44	6.64	5.14	1.92

was the need for reallocation of funds in favour of more bullock power, seed and other working capital and, less human labour and manure and fertilizer, in 1980-81; more human labour and bullock power and, less seed, manure and fertilizers, and other working capital in 2000-01 for optimizing the output. Such reallocation of funds could have contributed to increase in output by 3.50 per cent in 1980-81 and only 0.24 per cent, at the overall level, in 2000-01 over the existing level of output.

Amongst the zones, the estimates of the optimal resource use levels during the year 1980-81, in zone VI indicated the need to reallocate the available funds in favour of more human labour, manure and fertilizers, and other working capital and less seed; and in zone VII, by reducing the funds from human labour, bullock power, and reallocating more for seed and manure and fertilizers to obtain 2.63 and 3.17 per cent increase in the existing level of output. During the later period, however, there existed a very little potential i.e. below one per cent, in both the zones, and this little increase in output would be possible by reallocating funds in favour of more bullock power in both the zones and; more manure and fertilizers, in zone VI .

***Ratoon* sugarcane :** At the overall level, the difference between the optimal and existing use levels suggested that the funds should be reallocated in favour of more human labour, bullock power and manure and fertilizers to obtain 12.95 per cent increase in the output in 1980-81. Similarly, the reallocation of funds in favour of more human labour and bullock power, and less manure and fertilizers could increase existing level of output to the extent of 2.46 per cent in the year 2000-01.

The agro-climatic zonewise comparison between the optimal and existing levels of inputs suggested the need for higher use of human labour, manure and fertilizers, and other working capital and less use of bullock power in zone VI; less use of human labour and more use of bullock power and manure and fertilizers in zone VII to obtain 7.65 and 16.23 per cent increase in existing level of output during the year 1980-81. As against this, there existed the possibility to obtain 3.72 and 0.57 per cent increase in the existing output of respective zones, through reallocation of funds in favour of more human labour and other working capital and, less bullock power and manure and fertilizers in zone VI; and less bullock power and, manure and fertilizers, and more human labour in zone VII during the year 2000-01.

It is evident from the foregoing discussion that, under limited capital condition for production of crops, in general, the estimated optimal levels of human labour were lower than their existing levels for *kharif* paddy,

Table 5.27 : Comparison of Existing and Optimal Use Levels of Resource Inputs and Output of *Suru* Sugarcane

Particulars	*1980-81*			*2000-01*		
	VI	*VII*	*Overall*	*VI*	*VII*	*Overall*
Capital allocation (Rs.)	35555.63	38063.23	36413.52	25327.93	25593.98	25029.96
Human labour (mandays)						
Existing level	441.02	470.85	450.90	- -	- -	277.63
Optimal level	560.06	256.62	342.20	- -	- -	329.59
Bullock power (pairdays)						
Existing level	- -	25.55	23.59	15.27	13.67	13.65
Optimal level	-	16.20	36.82	74.18	60.07	32.08
Seed (kgs.)						
Existing level	7689.15	7635.48	7738.78	- -	- -	7214.51
Optimal level	6250.92	11292.41	7929.10	- -	- -	6134.40
Manures and fertilizers (Rs.)						
Existing level	14517.94	7099.80	12059.53	8550.94	- -	8788.74
Optimal level	20091.08	12918.01	3214.88	35963.53	- -	5418.09
Other working capital (Rs.)						
Existing level	7407.37	- -	7475.73	- -	- -	8687.56
Optimal level	11192.98	- -	12365.68	- -	- -	7159.17
Output (Qtl.)						
Existing level	943.42	854.75	914.03	981.76	996.78	989.17
Optimal level	968.27	881.86	946.02	990.83	997.92	991.51
Percentage increase	2.63	3.17	3.50	0.92	0.11	0.24

Table 5.28 : Comparison of Existing and Optimal Use Levels of Resource Inputs and Output of *Ratoon* Sugarcane

Particulars	*1980-81*			*2000-01*		
	VI	*VII*	*Overall*	*VI*	*VII*	*Overall*
Capital allocation (Rs.)	13550.68	17090.81	15298.27	14396.94	13166.94	13483.42
Human labour (mandays)						
Existing level	254.17	308.54	278.48	273.23	225.10	240.07
Optimal level	390.62	301.95	348.33	329.65	453.45	391.84
Bullock power (pairdays)						
Existing level	5.09	5.86	5.53	3.90	7.07	5.70
Optimal level	1.12	11.67	7.88	2.84	3.46	6.06
Manures and fertilizers (Rs.)						
Existing level	5742.81	4080.94	4999.85	6192.61	5396.14	5640.02
Optimal level	10716.27	14799.15	9345.03	3017.14	1737.12	4554.16
Other working capital (Rs.)						
Existing level	7450.27	- -	- -	4921.89	- -	- -
Optimal level	7622.54	- -	- -	5489.91	- -	- -
Output (Qtl.)						
Existing level	768.76	718.91	746.47	795.29	781.16	785.49
Optimal level	827.56	835.60	843.14	824.89	785.60	804.82
Percentage increase	7.65	16.23	12.95	3.72	0.57	2.46

rabi sorghum and *suru* sugarcane in 1980-81, and for *kharif* sorghum, *kharif* paddy and pearlmillet in 2000-01. The existing use levels of human labour were less than optimum during the year 1980-81 in the production of *kharif* sorghum, pearlmillet, groundnut, cotton, wheat and *ratoon* sugarcane, while that for groundnut, cotton, rabi sorghum, wheat, *suru* sugarcane and *ratoon* sugarcane in 2000-01. The existing levels of bullock power were found to be higher than the optimal levels in the production of *kharif* paddy, pearlmillet and cotton in 1980-81, and that for *kharif* paddy, groundnut, cotton and wheat in 2000-01. The estimated use levels of bullock power were less than optimum in the production of *kharif* sorghum, *rabi* sorghum, *suru* sugarcane and *ratoon* sugarcane in 1980-81 and that for *kharif* sorghum, peralmillet, *suru* sugarcane and *ratoon* sugarcane in 2000-01. The estimated levels of seed were lower than their optimal levels for *kharif* paddy, pearlmillet and *suru* sugarcane in 1980-81, that for cotton and wheat in 2000-01. While, the same were higher than optimum in the production of *kharf* paddy, groundnut and *suru* sugarcane in 2000-01. In the case of manure and fertilizers input, the optimal levels are lower than the existing levels in the production of *kharif* sorghum, cotton and *suru* sugarcane in 1980-81, and that for *kharif* sorghum, *kharif* paddy, pearlmillet, cotton, *suru* sugarcane and *ratoon* sugarcane in 2000-01. The optimal levels of other working capital were found to lower than their existing levels in the production of cotton and wheat in 1980-81, and that for *kharif* sorghum, *kharif* paddy, groundnut, cotton, wheat and *suru* sugarcane in 2000-01. The optimum levels of this input were higher than their existing use levels in the production of *kharif* sorghum, *kharif* paddy, groundnut, *rabi* sorghum and *suru* sugarcane in 1980-81 and that for pearlmillet in 2000-01.

The comparison of existing and optimum levels of resource inputs in the production of major crops indicated that the existing use levels of human labour, bullock power, seed, manure and fertilizer and other working capital were higher than the optimal levels in 3, 3, 0, 3 and 2 cases in 1980-81 and 4, 4, 2, 7 and 4 cases in 2000-01, at the overall level, respectively. Across the different agro-climatic zones, the same were higher from their optimum levels in 3, 5, 1, 7 and 4 cases during the year 1980-81 and in 7, 9, 3, 8 and 0 cases during the year 2000-01, respectively. In other words, the use of human labour, bullock power and seed was found excess at the overall level, as well as in the agro-climatic zones, during the later period, whereas, the use of manure and fertilizers and other working capital was found sub-optimal in the majority of crops, during the later period. The reallocation of limited capital in favour of more in these resource inputs is, therefore, advisable for all crops grown in different agro-climatic zones to optimize output from crop production.

5.6 Share of Technology in Crop Productivity

It has been realized from the foregoing discussion that during the period under reference agriculture in the State has shown changes in resource use structure, resource productivities and farmers' allocation efficiency. Adoption of crop production technology, wherever possible, and intensive cultivation have resulted in improvements in the productivities of individual resources. The importance of the individual resources has also undergone a change as a result of technological transformation of agriculture in different agro-climatic zones. The total change in production and change in productivity are two important dimensions of agricultural growth. Technological change is one of the major factors leading to changes in output. The adoption of new technology may affect the relative economic position of the factors of production in terms of their share in total output.

As transforming agriculture necessarily brings about changes in relative factor shares, this section attempts to decompose the total change in per hectare crop output with the effect of crop production technology into the proportion due to technical change and the proportion due to the change in the input use levels. The convenient economic model for decomposing total change in output is the production function. For any production function, the total change in output is brought about by shifts in the parameters that define the function itself and by changes in the volume of the inputs. The decomposition or 'component' analysis is the mathematical technique used to partition an aggregate into its component elements. Solow (1957) developed this approach to evaluate the effect of technological change on output growth in United State's agriculture. Bisaliah (1977) extended framework of decomposition analysis. The decomposition analysis is used to evaluate differences between new technology and old technology, with the hypothesis that, (i) there is a difference in the response function of new and old crop production process, the intercepts and slope coefficients differ, (ii) There is structural difference due to change in the efficiency parameters (intercepts) of the production function or changes in the slope coefficients of the input used, and (iii) or both. Hence, in the present investigation, the decomposition model (Bisaliah, 1977 and Kiresur et al., 1996) is used to decompose the total productivity differentials between two points of time, i.e. 1980-81 and 2000-01. The crop production technology during the year 1980-81 was treated as 'old' technology and the technology adopted during the year 2000-01 was treated as 'new' technology.

To sort out the contributions of technology and resource use differentials to the total productivity difference between 'old' and 'new' period, the log-linear production function of the type, specified in Chapter 3, was considered. In order to test the homogeneity of the production functions, the Chow test was applied to the two data sets i.e. 1980-81 and 2000-01, for individual crops at the overall level. The F-ratios estimated by Chow test are presented in Appendix- X. It was found that in both the periods, Chow's F-statistic was found highly significant, indicating that the parameters of the production function generated by 'new' period technology were different from those generated by 'old' period technology. These differences were due to changes in slope parameter as well as due to shifts in the intercepts. Hence, it can be concluded that, there was co-existence of neutral as well as non-neutral technological change in the production of major crops under study. This result support the rationale for decomposing the total change in output, which suggests that the decomposing analysis could be under taken with Cobb- Douglas per hectare production function.

With the decomposition equation (shown in Chapter 3), using the values of production parameters (Appendix- IX) and the input levels (Appendix- XI), the total change in per hectare output with the crop production technology have been decomposed. The results are presented in Table: 5.29 (A) to 5.29 (C). The percentage increment in the crop productivity due to technological change and increase in input use levels is depicted in Chart- 30.

***Kharif* sorghum :** A perusal of the results of decomposition analysis revealed that in case of *kharif* sorghum, there was slight discrepancy between the observed (77.59 %) and the estimated (78.81 %) difference in the productivities of the new and old technology. This discrepancy was attributed to the random error term, which accounts for variables that could not be included in the model such as management inputs. Such discrepancies of varying degree in decomposition analysis were also encountered in several earlier studies including those by Umesh and Bisalaih (1986), Ballabh and Sharma (1989), Lalwani (1990) and Kiresur et al (1996). In majority of these studies, such discrepancies were attributed to random errors and exclusion of one of the important variables from the model, namely, management inputs. However, in the present study, since the discrepancy in question was of very low order, the results of the decomposition analysis were considered to be satisfactory. (Badal and Singh, 2001)

Chart-30: Share of technology in the productivity of crops during the periods 1980-91 to 2000-01

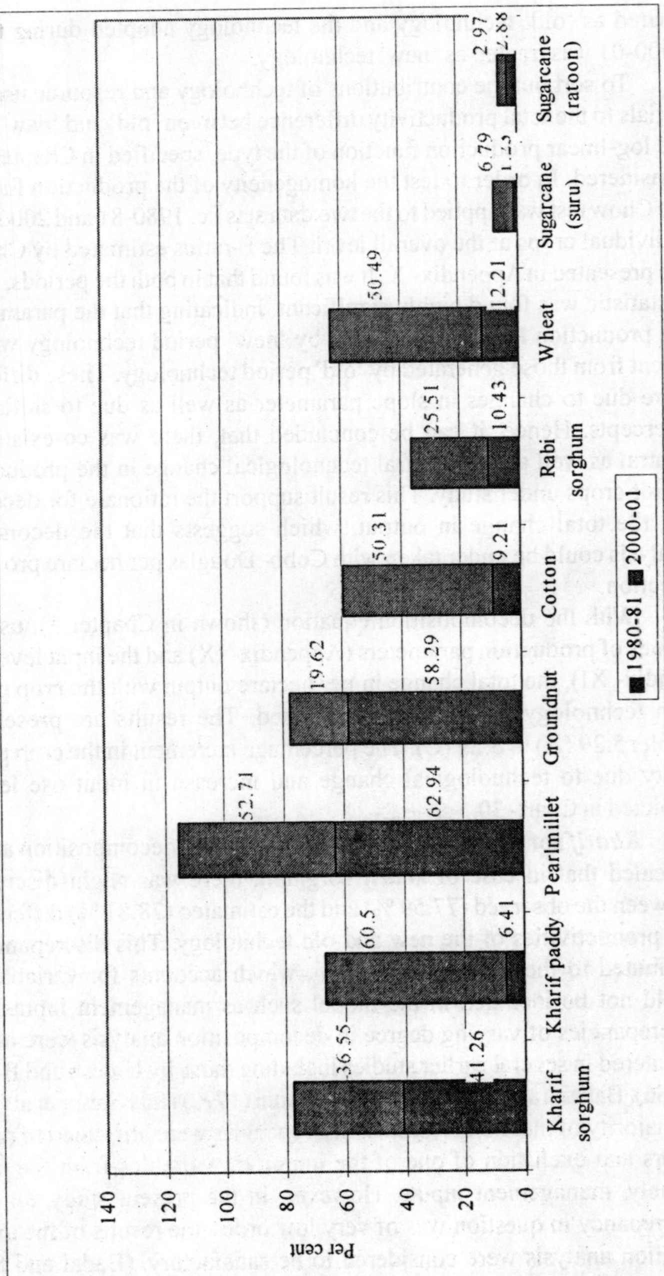

The technological and input use differentials between the two technologies together contributed to form the total productivity differences of the order of 77.81 per cent, whereas the technological component alone accounted for 41.26 per cent. This implies that with the present level of resource use during the year 1980-81, the productivity could be increased by 41.26 per cent, if the farmers could just switch over from old to new technology. Such an increase in the productivity exclusively due to technological improvement was brought about through a shift in the scale and/ or slope parameters of the production function. If, however, farmers could simultaneously raise the input use, as was in the year 1980-81 to the same level as in the year 2000-01, the productivity could be further increased by another 36.55 per cent to make a total productivity difference of 77.81 per cent.

The contribution of the neutral technological component in the productivity difference was positive and it was low order (11.37%) whereas, the non-neutral technological component contributed positively to the total difference in the gross value of output. This was to the tune of 29.89 per cent. The high positive neutral technological component signifies that there was improvement in the net efficiency of all the inputs taken together.

However, it was observed that with shift from old-period technology to new-period technology, there was a decrease in efficiency of human labour and seed making negative contributions to the productivity gap. The negative contribution of human labour and seed in the productivity gap indicates that the productivity gap was drastically reduced due to decrease in the use efficiency of these two inputs.

At the present level of input use (i.e. in 2000-01), a given increase in the use of these two inputs on both new and old farms could increase the productivity on the old farms more rapidly than that on the new farms. Therefore, the resource use efficiency on the new farms should be increased by way of adopting the package of cultivation practices in order to fully exploit the potential of new technology. On the other hand, the elasticities of bullock power, manure and fertilizers, and other working capital had positive contribution to the productivity difference, which was estimated to be nearly 40 per cent. These inputs together could substantially widen the productivity gap between the two technologies, as was quite expected.

An examination of the contributions made by different inputs in the total productivity gap between the two technologies revealed that manure and fertilizers caused the biggest jump in the productivity of the *kharif* sorghum during the year 2000-01. Next to this, other working capital was the

Table 5.29 (A) : Decomposition of Total Productivity Differences Due to Technology in *Kharif* sorhum, kharif paddy and pearlmillet

Sr No.	Source of productivity difference	Wheat: Percentage contribution Sub- total	Wheat: Total	Suru sugarcane: Percentage contribution Sub- total	Suru sugarcane: Total	Ratoon sugarcane: Percentage contribution Sub- total	Ratoon sugarcane: Total
A	Total difference observed in productivity		62.51		7.90		5.10
B	Due to difference in technology		12.21		3.55		2.88
1	Neutral technological difference	8.89		13.06		2.05	
2	Non-neutral technological difference	3.32		-9.51		0.83	
a	Human labour	1.01		39.60		41.43	
b	Bullock power	-2.23		4.46		-0.24	
c	Seed	0.37		-28.21		--	
d	Manure & fertilizers	3.37		-1.08		-33.63	
e	Other working capital	0.80		-24.28		-6.73	
C	Due to difference in input use level		50.49		4.61		2.97
a	Human labour	12.31		-12.98		-4.30	
b	Bullock power	-1.02		-4.45		0.15	
c	Seed	-0.43		-0.68		--	
d	Manure & fertilizers	21.62		8.86		9.20	
e	Other working capital	18.00		13.86		-2.07	
D	Total estimated difference in productivity (due to all sources)		62.69		8.16		5.85

Table 5.29 (B) : Decomposition of Total Productivity Differences Due to Technology Between 1980-81 and 2000-01

Sr No.	*Source of Productivity Difference*	*Groundnut*		*Cotton*		*Rabi sorghum*	
		Percentage Contribution		*Percentage Contribution*		*Percentage contribution*	
		Sub-total	*Total*	*Sub-total*	*Total*	*Sub-total*	*Total*
A	Total difference observed in productivity		76.25		57.80		34.87
B	Due to difference in technology		58.29		9.21		7.17
	1. Neutral technological difference	-79.62		72.25		93.23	
	2. Non-neutral technological difference	137.91		-63.04		-86.06	
	(a) Human labour	-42.28		-89.25		13.61	
	(b) Bullock power	158.85		-4.41		1.92	
	(c) Seed	145.66		36.61		0.27	
	(d) Manure and fertilizers	-22.96		-15.11		9.10	
	(e) Other working capital	-100.77		9.12		-110.96	
C	Due to difference in input use level		19.62		50.31		27.88
	(a) Human labour	5.67		24.42		29.93	
	(b) Bullock power	-23.73		0.45		7.15	
	(c) Seed	-1.37		0.97		0.66	
	(d) Manure and fertilizers	35.28		10.23		22.05	
	(e) Other working capital	3.77		14.25		-31.91	
D	Total estimated difference in productivity (due to all sources)		77.91		59.53		35.05

Table 5.29 (C) : Decomposition of Total Productivity Differences Due to Technology Between 1980-81 and 2000-01

Sr No.	Source of Productivity Difference	Wheat		Suru sugarcane		Ratoon sugarcane	
		Percentage Contribution		*Percentage Contribution*		*Percentage contribution*	
		Sub-total	*Total*	*Sub total*	*Total*	*Sub-total*	*Total*
A	Total difference observed in productivity		62.51		7.90		5.10
B	Due to difference in technology		12.21		3.55		2.88
	1. Neutral technological difference	8.89		13.06		2.05	
	2. Non-neutral technological difference	3.32		-9.51		0.83	
	(a) Human labour	1.01		39.60		41.43	
	(b) Bullock power	-2.23		4.46		-0.24	
	(c) Seed	0.37		-28.21		---	
	(d) Manure and fertilizers	3.37		-1.08		-33.63	
	(e) Other working capital	0.80		-24.28		-6.73	
C	Due to difference in input use level		50.49		4.61		2.97
	(a) Human labour	12.31		-12.98		-4.30	
	(b) Bullock power	-1.02		-4.45		0.15	
	(c) Seed	-0.43		-0.68		---	
	(d) Manure and fertilizers	21.62		8.86		9.20	
	(e) Other working capital	18.00		13.86		-2.07	
D	Total estimated difference in productivity (due to all sources)		62.69		8.16		5.85

strongest factor. Bullock power and seed was over used during the later period, hence, it caused the negative contribution in the total productivity. The total contribution of production gap attributed to the differences in the levels of input use was found to be 36.55 per cent.

***Kharif* paddy :** The per hectare production of *kharif* paddy with new production technology was about 66.31 per cent higher than that with old technology. The contribution of the technical change to the total change in output was estimated to 6.41 per cent. This value was obtained by adding first and second bracketed expressions on the right hand side of decomposition equation (mentioned in Chapter 3). With the same level of per hectare inputs used in 1980-81, about 6 per cent more output could be obtained in new period.

An increased use of seed, manure and fertilizers, and other working capital in 2000-01, has contributed 60.50 per cent of the increased output. Of the estimated 66.92 per cent contribution, the contribution of manure and fertilizers (44.82 %) was the highest, followed by seed (24.68 %) and other working capital (8.00 %), whereas, human labour and bullock power have contributed negatively with the magnitude of -12.12 and -4.88 per cent, respectively.

Pearlmillet : The discrepancy between the observed and the estimated differences in the productivities of pearlmillet during new and old period was 114.18 per cent. The difference in the total productivity contributed by technological component accounted for 62.94 per cent. The contribution of neutral technological component as well as non-neutral technology component in the productivity differences was positive. To the estimated difference in input use level (52.71 %), the contribution of other working capital (41.61 %) was higher followed by human labour (20.37 %) and, manure and fertilizers (2.61 %).

Groundnut : The per hectare production of groundnut in 2000-01 was higher by 76.25 per cent than that in 1980-81. The contribution of technical change to total change in output was estimated to be 58.29 per cent. The contribution of differences in input use levels was 19.62 per cent, of which, human labour, manure and fertilizers, and other working capital have contributed by 5.67, 35.28 and 3.77 per cent, respectively. The use of inputs viz. bullock power and seed have, however, turned out to be negative. The estimated difference in the productivity of groundnut due to all sources accounted for 77.91 per cent.

Cotton : The per hectare production of cotton with new production technology was 57.80 per cent higher than that with old technology. The contribution of the technical change to the total change in output was estimated to 9.21 per cent. It means, with the same level of per hectare inputs used in 1980-81, about 9 per cent more output could be obtained in new period. An increased use of human labour, other working capital and, manure and fertilizers in 2000-01, have contributed about 50 per cent increase in output. The contribution of human labour, (24.42 %) was the highest, followed by other working capital (14.25 %) and manure and fertilizers (10.23 %).

***Rabi* sorghum :** The discrepancy between the observed (34.87 %) and the estimated differences (35.05 %) in the productivities of *rabi* sorghum during new and old period was about one per cent. The difference in the total productivity contributed by technological component accounted for 7.17 per cent. The contribution of neutral technological component in the productivity differences was positive (i.e. 93.23 %), whereas, non-neutral technology component contributed negatively by 86.06 per cent. To the estimated difference in the input use level (i.e. 27.88 per cent), the contribution of human labour (29.93 %) was the highest followed by manure and fertilizers (22.05 %) and bullock power (7.15 %).

Wheat : The per hectare production of wheat with new production technology was observed to be 62.51 per cent higher than that with old technology. The contribution of the technical change to the total change in output was estimated to be 12.21 per cent. With the same level of per hectare inputs used in 1980-81, the output could be obtained to the extent of 50.49 per cent in new period. In the increased use of resources, the contribution of manure and fertilizers (21.62 %) was the highest followed by other working capital (18.00 %) and human labour (12.31 %) in the estimated differentials due to increased input use, during the later period. The difference in the input use level was estimated to 50.49 per cent and total estimated difference in productivity due to all sources accounted for 62.69 per cent. The contribution of bullock power and seed was negative, in the estimated difference due to input use.

***Suru* sugarcane :** The per hectare production of *suru* suagarcane in 2000-01 was higher by 7.90 per cent than that in 1980-81. The contribution of technical change to total change in output was estimated to be only 3.55 per cent. The contribution of differences in input use levels was 4.61 per cent, of which, manure and fertilizers, and other working capital have contributed to 8.86 and 13.86 per cent share, respectively. The magnitude of use of inputs viz. human labour, bullock power and seed have turned out to be negative. The estimated difference in the productivity of *suru* sugarcane due to all sources accounted for 8.16 per cent.

***Ratoon* sugarcane :** The per hectare production of *ratoon* sugarcane with new production technology was 5.10 per cent higher than that with old technology. As like *suru* sugarcane, the contribution of the technical change to the total change in output was found very low (i.e. 2.88 %), indicating that with the same level of per hectare inputs used in 1980-81, about 3 per cent more output could be obtained in new period. An increased use of only manure and fertilizers in 2000-01, has contributed about 3 per cent of the increased output. The contribution of human labour and other working capital have turned out to be negative in the estimated difference due to input use level (2.97 %). The estimated difference in the productivity of *ratoon* sugarcane due to all sources accounted for 5.85 per cent. In both the types of sugarcane, the productivity was marginally increased and the share of technology in this increased.

Productivity was only 3 per cent. This indicated the weaknesses in the adoption and management of technology on farmer's field.

The decomposition analysis revealed that there were slight discrepancies between observed and the estimated differences in the productivities of new and old technology. This discrepancy was attributed to the random term, which accounts for variables that could not be included in the model such as, management inputs.

The technological and input use differentials between two technologies together contributed to form the total productivity differences of the order 35.55 to 77.91 per cent in *rabi* sorghum, cotton, wheat, *kharif* paddy, *kharif* sorghum and groundnut. It was the highest in pearlmillet (115.65 %) and the lowest in *ratoon* sugarcane (5.85 %). The technology component alone accounted for in the range of 2.88 per cent in *ratoon* sugarcane to 62.94 per cent in pearlmillet. This implies that with the level of resource used during the year 1980-81, the productivity could be increased by about this range. Such an increase in productivity was exclusively due to technological improvement brought through a shift in the scale and / or slope parameter of the production function. If however, the farmers could simultaneously raise the input use level to the same level as in the year 2000-01, the productivity could be further increased to the extent by 2.97 per cent in *ratoon* sugarcane to 60.50 per cent in *kharif* paddy.

The contribution of neutral technological component in the productivity difference were positive, except groundnut, which signifies that with the level of inputs used during the year 1980-81, the farmers could have trebled the productivity level by adopting new technology, provided the efficiency levels of input use held constant.

The non-neutral technological component found negative, except groundnut and wheat, in all the crops under study. The main factor responsible for the negative non-neutral technological component was human labour, contributing negatively in the total productivity difference. The negative contribution of human labour indicated that the productivity gap was drastically reduced due to the decrease in the use efficiency of this input.

Kiresur et al. (1995) estimated the total difference in productivity of sorghum in Karnataka to about 45 per cent. Of this, 35 per cent was contributed by technology and remaining 10 per cent was shared by additional input use. Sisodiya et al. (1999) also found that total output of pigeon-pea in Madhya Pardesh was increased by 59.73 per cent due to technology. The findings of this study confirm the study results of Kiresur et al. (1995) and Sisodiya et al. (1999). Hence, the hypothesis viz; the increased productivity found to be associated with the impact of technologies during the period under study, was proved.

5.7 Economic Sustainability of Crop Production Technology

The impact or the potential of any improved technology under real farm situation is generally assessed from the magnitude of the difference in the mean yields, net economic returns or benefit-cost ratio of the improved technology and those of old or traditional technology. In general, the yields or returns from new technology exceed those from old one (Kiresur et al, 1996). However, it is most important to know whether the additional yield or returns offered by new technology is sufficient enough to qualify the technology in terms of economic viability, as well as to know whether the technology guarantees sustained productivity across various agro-climatic zones and crop growing situations, and over the years in a given locations, for ready acceptance and adoption of the technology and its sustenance on the farmers' field, over a period underconsideration.

In the present investigation, a model that uses both the actual yield during the year 2000-01, due to new technology as well as its coefficients of regression on the site or environment indices, which, in turn represent the yield levels of old period i.e. 1980-81, and is examined and evaluated against a standard or hypothetical yield to assess the economic sustainability of a crop production technology during the period under study. The results on comparison of the actual yield levels during the year 1980-81 and sustainability index of technology adopted during the year 2000-01 along with the values of Desired yield (Di) and Standard yield (Ds), in respect of selected crops under study is depicted in Table 5.30.

Table : 5.30 Actual Yield Levels during 1980-81 and Sustainability Index of Technology adopted during the Year 2000-01 for crop Production

Crop	*Zone*	*Yield (qtls.)*		*Di (qtl./ha)*	*Ds (qtl./ha.)*	*SI (Di/Ds)×100*
Kharif sorghum	VII	15.30	22.95	27.28	30.27	90.12
	VIII	7.11	16.95	17.79	18.19	97.80
	Overall	9.33	19.57	24.15	27.05	89.28
Kharif paddy	II	9.44	20.34	21.32	24.53	86.91
	V	10.80	29.83	23.21	22.83	101.66
	IX	13.54	23.55	24.38	24.59	99.15
	Overall	11.75	22.95	23.10	24.35	94.87
Pearlmillet	V	2.60	16.00	17.54	17.96	97.66
	VI	4.21	9.46	11.70	11.91	98.24
	VII	6.44	9.11	9.58	9.57	100.10
	Overall	3.93	11.07	13.13	13.20	99.47
Groundnut	V	13.26	18.68	21.93	23.88	91.83
	VII	5.66	8.00	9.23	10.14	91.03
	Overall	8.91	12.19	18.83	20.45	92.08
Cotton	VII	3.44	7.56	7.44	7.78	95.63
	VIII	3.63	5.48	6.36	6.43	98.91
	Overall	3.52	6.25	6.66	7.46	89.28*
Rabi sorghum	VI	6.05	11.60	11.53	14.07	81.95
	VII	8.67	9.60	10.70	12.63	84.72*
	Overall	7.06	10.55	10.48	13.85	75.67*
Wheat	VI	15.10	22.95	25.18	26.32	95.67
	VII	17.83	19.44	17.62	17.33	101.67
	VIII	8.08	16.54	18.15	18.73	96.90
	Overall	14.32	18.79	19.95	20.31	98.23
Sugarcane *(Suru)*	VI	943.42	981.76	945.75	943.54	100.23
	VII	854.75	996.78	913.70	933.76	97.85
	Overall	914.03	989.17	934.23	940.90	99.29
Sugarcane *(Ratoon)*	VI	768.76	795.29	827.18	831.17	99.52
	VII	718.91	781.16	762.67	728.25	104.73
	Overall	746.47	785.49	805.44	764.47	105.36

Di = Yield level during 2000-01 adjusted for regression
Ds = Standard yield level (Mean yield in 2000-01+ Symmetry of site index)

SI = Sustainability index
Note : * indicates the Di values are significantly lower than Ds at 10 percent significance level)

The sustainability of the new technology could be reflected by its proportional response to the existing technology situation. The results presented in Table 6.30, on the computation of sustainability index (SI) for the new and old technology revealed that, at the overall level, the Desired yield level (Di) was found lower than Standard yield level (Ds) in all the crops under study, except in *ratoon* sugarcane. The sustainability index were found varied in the range between 75.67 to 105.36 per cent in terms of yield among the selected crop enterprises. Across the agro-climatic zones, the 'Di' values were found higher than the 'Ds' values for *kharif* paddy in zone V, wheat in zone VII, *suru* sugarcane in zone VI and *ratoon* sugarcane in zone VII. In the case of cotton and *rabi* sorghum, the sustainability index were found significantly lower i.e. 'Di' value of these crops was significantly lower than the 'Ds' value. This indicated the weakness in implementation and management of agricultural technology over the period under study. Thus, lower values of the sustainability index also suggested that there could be shortfalls in the implementation or application of the new crop production technology in a given situation. The analysis brought out that the magnitude of yield realized by crop production method adopted for all the crops under study, except cotton and *rabi* sorghum, during the year 2000-01, was found sustainable.

According to Caviglia (1992), the yield of sunflower was increased by 8 per cent. The farm holding using sustainable technology had monetary results 40 per cent higher than traditional holding. Kiresur et al. (2001) found that improved technologies registered 24 to 107 per cent higher yields over the traditional yield level of oilseeds in Karnataka. The crop production technology was found fairly sustainable as sustainability indices ranging from 15 to 100 per cent in terms of yield. The findings of this study found to be consistent with the study results of Kiresur (2001). Hence, the hypothesis viz; the crop production technologies found to be sustainable and can improve the economic benefits on the farmer's field, was proved.

6

Summary and Conclusions

A perceptible technological transformation has been underway in Indian agriculture since the mid-'sixties'. The new high yielding agricultural production technology has resulted in an upward shift in the production functions for major crops, disturbing thereby the stable long-run equilibrium of traditional agriculture. There are, however, some doubts regarding distribution of benefits of new agricultural technology over the farms and regions in India. The inter-regional and inter-zonal disparities in the adoption of new technology have proved that technological improvements certainly overcome many of the limitations imposed by natural conditions, but beyond a point, increase in productivity or removal of uncertainties becomes impossible.

The State Agricultural Universities play a vital role in the development of technologies appropriate for region, climate and needs of the farmers. These research institutions spend huge amounts on research and extension for evolving number of technologies. There are differentials in the agro-climatic features of agro-climatic zones. Moreover, there exists imbalance and disparities in the income levels of the farms. The crux of the problem of modernization of agriculture is associated with the availability of capital resources, efficiency in allocating these limited resources for different uses, nature of technology and extent of its adoption.

Maharashtra state is one of the several regions in the Indian Union, where technological advances in agriculture have not made much headway in their contribution to increased agricultural productivity and production. There are doubts regarding the transfer of the benefits of new agricultural technology in Maharashtra, which has a wide range of highly variable agro-climatic factors. The State has variety of climatic patterns. The vast area of 30.8 million hectares with varying soils, topography, water resources and climatic features is, therefore, bound to have different agro-climatic conditions. The nine agro-climatic zones have been identified in the State on the basis of varying rainfall pattern, topography, soil characteristics, climate and cropping pattern.

National Agricultural Research Programme (NARP) have helped to develop linkage between Research and Extension. The programme was initiated in the year 1984 and efforts were being made to develop location specific technologies based on agro-climatic conditions. It was decided to cover a comprehensive study on impact of technology on productivity of crops due to above programmes. The present study viz; ***Impact of Technology on Productivity of Major Crops in Maharashtr*** was taken up for the two points of time, i.e. 1980-81 and 2000-01, in order to understand new production functions with prime concern for increasing resource use within a given technological environment. An attempt was made to examine the impact of technology on productivity of major crops and the pattern of resources use structure, resource productivities, and allocation efficiencies in the major crops in the different agro-climatic zones of Maharashtra.

The study was undertaken with the following specific objectives.

(i) To assess the level of adoption of crop production technology in different agro-climatic zones.

(ii) To find out technological impact on use of crucial inputs and productivity of major crops in different agro-climatic zones.

(iii) To study the changes in costs and returns structure of major crops over a period of time.

(iv) To examine the changes in resource use efficiency and determine the optimum use levels in production of major crops.

(v) To find out the factor share of the technology in the productivity of major crops over a period of time.

(vi) To assess the economic sustainability of crop production technology over a period of time.

The study is of importance to recognize whether the new agricultural technology is profitable. The findings would be useful to the farmers, researchers and planners to understand the benefits of the research and extension efforts made under NARP. Moreover, the findings of the study would help the planners and decision makers to formulate effective strategies for development of agriculture in other agro-climatic zones. The main purpose of the study was to understand the growth of agriculture in different agro-climatic zones by way of estimating changes in resource use structure, resource productivities and allocation efficiency on farms during pre-technological and post-technological change periods. The implementation of NARP during the year 1983-84 to 1996-97 was the yardstick for pre-technological and post-technological period. The year 1980-81 and 2000-01 (being the normal years) were treated as pre-technological and post-technological periods, respectively. The investigation was based on the cross-sectional farm level data obtained under the Government of Maharashtra sponsored Cost of Cultivation Scheme by cost accounting method from the sample farms in different clusters of villages falling in different agro-climatic zones of Maharashtra for the two points of time. For sugarcane crop, however, the data collected through the Government of India sponsored Comprehensive Scheme for Cost of Cultivation of Principal Crops were used as no other source of data was available. In all for nine crops, 397 sample farms were selected for each of the periods, i.e. pre-technological and post-technological period. The precaution was taken that agro-ecological and socio-economical features remain the same for sample farms selected in 2000-01, so as to avoid the differences in the socio-economic conditions within the sample farms selected during the base year. Under both the cost of cultivation schemes, the samples were selected by adopting three-stage-stratified random sampling plan with homogenous crop growing zones as strata, tahsil as primary unit, a cluster of three villages as the secondary unit and an operational holding within the cluster as the ultimate unit of sampling.

Based on the farm level data for the agricultural years 1980-81 and 2000-01, attempts were made to evaluate the changes in cropping pattern, cropping intensity and resource endowment in different agro-climatic zones. The changes in the structure of resource use, farm income, elasticity of production of farm resources and allocative efficiency and scope for reallocation of farm resources for the crop production business were studied. Finally, the study addressed itself to the relative factor shares of technology in the productivity of crops that have occurred during the period as a result of technological change in agriculture in different agro-climatic zones.

Primarily, the tabular analysis based on simple means and percentages has been used in this study to indicate the level of technology adoption, resource use levels, costs and returns structure for selected crop production enterprises. The costs and returns data corresponding to the first time period have been inflated to the 2000-01(second time period) prices in order to overcome the problem of price rise. The Cobb-Douglas type of production function was fitted to the data on the output of crops as dependent variable and human labour (mandays), bullock power (pairdays), seed (Kg.), manure and fertilizers (Rs.), other working capital (Rs.) and technology adoption index (per cent) as independent variables for individual crop grown in different agro-climatic zones. The production function estimates were deployed to estimate resource productivities and returns to scale underlying in the crop production processes in different agro-climatic zones. The same function has been again attempted by excluding a variable viz. technology adoption index, to show changes in resources productivities and allocation efficiencies. The use of decomposition model has been made to study the factor share of technology in the productivity of the crops under study and the sustainability of the crop production technology was judged by the sustainability index.

6.1 Summary

The major findings of the study are summarized below :

1. The technology adoption index over a period of 20 years was found increased to the extent of 13.86 per cent in groundnut to 245.85 per cent in pearlmillet.
2. The magnitude of the technology adoption index during the post technology period was in the range between 24.12 and 49.84 per cent.
3. The estimated adoption index was higher in all the agro-climatic zones during the year 2000-01 than that of the year 1980-81. Among the different agro-climate zones, no definite trend was observed.
4. At the overall level, during the year 1980-81, the magnitude of composite indices were found to be in the range between 8.74 to 23.18 per cent, 21.82 to 43.59 per cent and 65.27 to 71.83 per cent under low, medium and high level category of the technology adoption, respectively. While, the figures for the same under respective categories were 17.86 to 26.47 per cent, 37.88 to 53.64 per cent and 71.83 to 77.26 per cent during the year 2000-01. The increase in the magnitude of the composite index was because of impact of technology that changed the adoption behaviour of the farmers.
5. Human labour required to produce per hectare of *kharif* sorghum, pearlmillet, cotton, *rabi* sorghum and wheat increased considerably during the period, while for *kharif* paddy, *suru* sugarcane and *ratoon* sugarcane, it declined. In the case of

groundnut, per hectare use of human labour remained more or less the same during the period.

6. The per hectare use of bullock power declined substantially for all the crop enterprises, except cotton. In the case of cotton, per hectare use of bullock power has marginally increased (4.73 % only). The reason for the decline in the use of bullock power might be due to substitution by mechanical power.
7. The per hectare use of seed declined during the period by about 1.72 to 11.81 per cent for *kharif* sorghum, groundnut, cotton, wheat and *suru* sugarcane. This was only because of improvement in the management practices and adoption of high yielding varieties. However, the per hectare use of seed increased for *kharif* paddy, pearlmillet and *rabi* sorghum. Generally, these crops are grown in light soils under rainfed conditions and for obtaining higher fodder production required to feed the cattle, it is the tendency of the farmer to use more seed.
8. The use of farm yard manure has been observed to be relatively more in all the crops under study, except sugarcane. By and large, the cattle population was more than doubled during the period. With the increase in cattle population, the availability of FYM also increased. However, still there was a wide gap between existing and recommended use levels of manure. The sugarcane growers met their need of manure with substitution of inorganic fertilizers, hence, the decline in the use of manure for producing sugarcane was observed.
9. In the base year 1980-81, the use of fertilizers was at lower level. During the later period, due to the better linkage of research and extension, the awareness about fertilizer use increased, as a result, there has been an increasing tendency to use fertilizers. However, the quantum of the fertilizers used for different crop enterprises was far lower than the recommended doses. There was variability in allocating the fertilizer resource among different agro-climatic zones.
10. Except seed and bullock power, the per hectare use of all other resources was relatively higher for the crop enterprises produced in different agro-climate zones during the period. The increase in the use of human labour, manure and fertilizers and other working capital could be attributed to the improved technology management practices.
11. The productivity of all the crops increased during the period.

However, in the case of sugarcane, the productivity increased only marginally. (i.e. by 5 to 8 %).

12. The per hectare cost of cultivation of all the crops increased by 3.03 to 90.35 per cent, during the period, except groundnut, wheat and sugarcane. The lowest and the highest being, respectively, for *kharif* sorghum and *rabi* sorghum. The per hectare cost of cultivation, in fact, declined by about 5 to 21 per cent in the case of sugarcane during the period due to the more use of machine power for the different farm operations.
13. The gross returns obtained from different crop enterprises increased during the period for all the crops under study, whereas in the case of sugarcane, it remained more or less unchanged. The increase in gross returns was due to increase in the use of resources for obtaining the higher yields through adoption of the improved crop production technology.
14. The various measures of the farm efficiency showed significant improvement in the efficiency in managing the crop production activities. The value of gross output per manday increased in all the crops by about 20 to 260.54 per cent, whereas it declined in *rabi* sorghum and wheat. It remained more or less the same in the case of cotton. The gross returns per manday and paid-out cost per unit of output got reduced in *rabi* sorghum and wheat. This might be due to the lowering down of the farm-labour efficiency on these farms. The total cost per unit of output declined in the range of 9.94 to 54.95 per cent in all the crops, except *rabi* sorghum. In the case of *rabi* sorghum, the total cost per unit of output increased by 27.38 per cent. Similarly, the various ratios in respect of c sts and output indicated definite sign of improvement in the farm efficiency due to the improved management practices and technology adoption.
15. With the help of input saved approach, the value of resources saved with the introduction of new technology and the quantity of extra output produced during the period was estimated. At the overall level, the value of input saved due to new crop production technology was observed to be the highest in cotton (Rs. 4.65), while the same was the lowest in *ratoon* sugarcane (Re. 0.05). However, in the case of *rabi* sorghum, there was no saving.
16. The quantity of extra output obtained with new technology using the old technology volume of inputs were estimated. The quantity of extra output due to new crop production technologies, was

estimated in the range between 0.44 quintal in cotton to 19.66 quintals in *suru* sugarcane, during the period under study, except *kharif* paddy in zone IX and *rabi* sorghum in zones VI and VII, wherein it was found negative.

17. The production elasticities of human labour, bullock power, and seed declined. The decrease in the production elasticities in human labour, bullock power and seed showed the relative importance of human labour and bullock power would decline in the process of transformation of agriculture from traditional to modern methods of crop production. On the other hand, the production elasticities of manure and fertilizers, other working capital (which includes the expenditure on machine charges, irrigation, plant protection measures, land revenue and other taxes, repairs to the farm implements, incidental charges etc.) and technology adoption index increased to some extent during the period. The increase in the magnitudes of the production elasticities of these resource variables during the period supported the hypothesis of positive impact of the technological change during the period. It also indicated that, there was a scope for intensifying applications of these inputs for agricultural production in the zones.
18. The production elasticities of all the input variables, with few exceptions, were less than unity during both the years, implying decreasing marginal productivity of factor inputs. This means that if other inputs are fixed at a specified level and only one input is increased, the gross returns will increase at a diminishing rate.
19. The comparison of intercept term i.e. constant (Log_{10} a) revealed that there has been technological change in production of majority crop enterprises, except wheat and *ratoon* sugarcane, as the magnitude of intercept term was higher during the later year over the base year. The degree of technological change, however, varied considerably among the crops as well as among the agro-climatic zones.
20. The comparison of the sums of production elasticities of the individual crops among the different agro-climatic zones indicated constant returns to scale in production of the all the crops under study during both the time periods.
21. The marginal value product of human labour was found significantly higher than the wage rates in cotton during the year

1980-81 and in groundnut, cotton and *ratoon* sugarcane during the year 2000-01.

22. The use of bullock power was made sub-optimally in the production of *kharif* paddy in zones II and IX, pearlmillet in zone VI, groundnut in zone V, *rabi* sorghum in zones VI and VII and *ratoon* sugarcane in zone VII in the year 1980-81. While, the use of this input was found to be inefficient in the production of *kharif* paddy in zone V and IX, groundnut in zone VII, cotton in zone VII, wheat in zone VII and VIII and *ratoon* sugarcane in zone VI during the year 2000-01.
23. The greater level of resource use inefficiency existed in the case of seed in the production of *kharif* paddy in all the zones and pearlmillet in zone VI during the year 1980-81, whereas the use of seed was found sub-optimal in the production of *kharif* sorghum in zone VII, *kharif* paddy in zones V and IX, cotton in zone VII and *rabi* sorghum in zone VI during the year 2000-01.
24. The manure and fertilizer was used sub-optimally during the year 1980-81 in the production of pearlmillet in zone V, *rabi* sorghum in zone VII, whereas it was used rationally in the production of all crops during the year 2000-01.
25. The use of other working capital was found to be rational in the production of majority of crops, during both the periods. However, during the year 1980-81, it was sub-optimally used in the production of *kharif* paddy in zone IX and groundnut in zone VII. This indicated the improvement in the management practices, due to the awareness of the technology.
26. The estimated optimal use levels of resources under limited capital condition indicated that, the productivity of the crops could have been increased by reallocating the available resources as follows.

Resource Use	*More Scope*		*No Scope*	
	1980-81	*2000-01*	*1980-81*	*2000-01*
Human labour	Kh. sorghun, Pearlmillet, Groundnut, Cotton, Wheat & Ratoon sugarcane	Groundnut, Cotton, Rabi sorghum, Wheat, Suru and Ratoon Sugarcane	Kh. paddy, Rabi sorghum and suru sugarcane	Kh. sorghum Kh.paddy and Pearlmillet

(Contd.)

(Contd. Table)

Resource Use	*More Scope*		*No Scope*	
	1980-81	*2000-01*	*1980-81*	*2000-01*
Bullock power	Kh. sorghum, Suru & Ratoon sugarcane	Kh. sorghum, Pearlmillet, Suru and Ratoon sugarcane	Kh.paddy, Pearlmillet and Cotton	Kh. paddy, groundnut, Cotton and Wheat
Seed	Kh. paddy pearlmillet and Suru sugarcane	Cotton and Wheat		Kh. paddy, Groundnut and Suru sugarcane
Manure and Fertilizers	Kh. paddy, Groundnut, Wheat and Ratoon sugar-cane	Groundnut, Rabi sorghum and Wheat	Kh. sorghum, Cotton and Suru sugarcane	Kh. sorghum, Kh. paddy, Pearlmillet, Cotton, Suru and Ratoon sugarcane
Other working capital	Kh. sorghum, Kh. paddy, Groundnut, Rabi sorghum and Suru sugarcane	Pearlmillet	Cotton and Wheat	Kh. sorghum, Kh. paddy, Groundnut, Cotton, Wheat and Suru sugacane

27. The technological and input use differentials between old and new period technologies together contributed to the total productivity differences, which were the lowest in *ratoon* sugarcane (5.85 %) and the highest in pearlmillet (115.65 %).
28. The technology component alone accounted for 2.88 to 62.94 per cent in the production of crops under study. This implies that with the level of resource used during the year 1980-81, the productivity could be increased in this range.
29. The contribution of neutral technological component in the productivity differences were positive, except in groundnut, which signifies that with the level of inputs used during the year 1980-81, the farmers could have increased the productivity level by adopting new technology, provided the efficiency levels of input use are held constant.
30. The non-neutral technological component was found negative in *kharif* paddy, cotton, *rabi* sorghum and *suru* sugarcane,

indicating the reduction in contribution of total technological component in the productivity. The main factor responsible for the negative non-neutral technological component was the excess use of human labour, contributing negatively in the total productivity difference.

31. The computation of sustainability index (SI) for the new and old technology revealed that the desired yield level (Di) was lower than standard yield level (Ds) in all the crops under study, except *ratoon* sugarcane.
32. Among the different agro-climatic zones, the 'Di' values were found higher than the 'Ds' values for *kharif* paddy in zone V, *suru* sugarcane in zone VI and wheat and *ratoon* sugarcane in zone VII.
33. The sustainability index varied from 75.67 to 105.36 per cent in terms of yield among the selected crop enterprises.

6.2 Conclusions

In the light of the empirical evidence brought out by the study, the following conclusions have been drawn :

1. The level of adoption of the technology increased considerably during last 20 years (1980-81 to 2000-01). Amongst the five technology parameters, the seed rate occupied the maximum score. Farmers gave least importance to the plant protection measures, except in the case of cotton. The composite index of technology was higher in all the agro-climatic zones during the year 2000-01 than that in the year 1980-81. It was important to note that the magnitude of the composite index of technology adoption was less than 50 per cent, indicating thereby the constraints in adoption of new technology and suggesting measures for removal of those constraints.
2. The analysis regarding the resource-mix employed on the sample farms for both the years indicated that there have been structural adjustments in the pattern of resource use on the farms. The human labour requirement increased in majority of the crops. The change in adoption of intensive cultivation practices has resulted in an upward shift in the demand for the human labour. For bullock power, the per hectare use declined substantially for all the crop enterprises, except cotton. This decline was due to substitution by mechanical power i.e. tractor, thresher etc. The use of seed declined by 1.72 to 11.81 per cent for *kharif* sorghum,

groundnut, cotton, wheat and *suru* sugarcane, during the period. The decline in the use of seed was mainly attributed to the adoption of high yielding varieties. However, the per hectare use of seed rate increased in the crops viz; *rabi* sorghum, *kharif* paddy and pearlmillet. This was due to a tendency of the farmers to use a higher seed rate for the cereal crops such as *rabi* sorghum and pearlmillet grown in the light and medium soils under rainfed conditions. The use of farm-yard manure has been observed to be relatively more than the use in base period (i.e. 1980-81) in all the crops under study, except sugarcane. The low use of manure in sugarcane was due to limitations in the supply of farm-yard manure and the substitution by inorganic fertilizers. Concerning to the fertilizer resource, it was seen that during the year 1980-81, the use of this input was already at lower level. Due to the better linkage, the awareness increased, as a result, there was an increasing tendency to use fertilizers. The great variability was observed in allocating the fertilizer resource among different agro-climatic zones. The allocation of fertilizer resource among different agro-climate zones seemed to have governed more by productivity and output price expectations, rather than any other considerations.

3. The per hectare output of crops have increased considerably during the period under study. The rate of increase, however, varied among the crop enterprises.
4. The structural adjustments in the use of farm resources have an economic impact on the costs and returns of the individual crop enterprises of the sample farms. The per hectare cost of cultivation of all the crops, except sugarcane, increased by 3 to 90 per cent during the period. The relative increase in the cost of cultivation of different crop enterprises was because of increased use of different farm resources, which resulted in increase in variable cost for obtaining the higher yields through adoption of the new technology. Surprisingly, the cost of cultivation of sugarcane (i.e. *suru* and *ratoon* cane) declined over a period of time, indicating that the sugarcane production management technology found to be stagnant and it needs a breakthrough for increasing the productivity of sugarcane.
5. The gross returns obtained from different crop enterprises have undergone considerable increase in response to structural adjustment in the use of farm resources, during the period under study. The per hectare gross returns obtained from *suru* sugarcane as well as *ratoon* sugarcane remained more or less unchanged.

6. The cost per unit of output of crops declined by 9.40 to 54.95 per cent during the period under study. The reduction in cost per unit of output could be attributed to the increase in the productivity of the crops due to impact of new technology.
7. The changes in resource use structure, costs and returns have indicated that agriculture in the agro-climatic zones has been on the move from its traditional stage to modernization. However, greater dependence on rainfall (which is generally scanty and unevenly distributed) for cultivation of foodgrains and other crops has proved to be a major constraint in the process of modernization of agriculture in the State.
8. The examination of the production function for the individual crop enterprises on the sample farms in 1980-81 and 2000-01 revealed that the production elasticities of human labour, bullock power, and seed have declined, while those of manure and fertilizers, and other working capital increased during the period. The decrease in the production elasticities of human labour, bullock power and seed clearly indicated that the relative importance of these inputs declined in the process of modernization of agriculture. On the other hand, higher production elasticities of non-conventional inputs i.e. manure and fertilizers, other working capital and technology adoption level during the later period have clearly shown that in the process of transformation of agriculture in the State, there was higher premium on investment in technological changes.
9. There prevailed constant returns to scale in the production of the individual crops among the different agro-climatic zones during both the time periods.
10. The comparison of marginal value products of the resource variables in the production of major crops indicated that the marginal value products of inputs significantly differed from their unit prices. This suggested that the degree of resource allocation efficiency was relatively higher in 2000-01 than that in 1980-81. This particular finding was a clear indication of the fact that since agriculture in the State was in the transitional phase of development it was a state of disequilibrium in the pre-technological period and it might take some time to attain new equilibrium at a higher level.
11. The comparison of existing and optimum levels of resource inputs in the production of major crops under limited capital constraint, indicated that the use of human labour, bullock power and seed

was found in excess during the later period, whereas the use of manure and fertilizers and other working capital was found sub-optimal for majority of the crops, during the later period. The reallocation of available capital in favour of more for these resource inputs, is therefore, advisable for all the crops grown in different agro-climatic zones to optimize output from crop production. However, the use levels of these resources increased during the period. It suggested that these resources were not managed well. There was a need to create the awareness among the farmers regarding the timely operations and management practices for harvesting the benefits of the improved technology.

12. The decomposition analysis revealed that the technology component alone accounted for in the range of 2.88 per cent in *ratoon* sugarcane to 64.94 per cent in pearlmillet. This implies that with the level of resource used during the year 1980-81, the productivity could be increased in this range. Such an increase in the productivity exclusively due to technological improvement is brought through a shift in the scale and/or slope parameter of the production function. The contribution of neutral technological component in the productivity difference was positive in majority of the crop enterprises, which signifies that with the level of inputs used during the year 1980-81, the farmers could have increased the productivity levels by adopting new technology. The non-neutral technological component was found negative, in the crops viz; *kharif* paddy, cotton and *rabi* sorghum. The main factor responsible for the negative non-neutral technological component was human labour, contributing negatively in the total productivity difference. The negative contribution of human labour indicated that the productivity gap drastically got reduced due to the decrease in the use efficiency of this input.
13. The sustainability indices in the case of cotton and *rabi* sorghum were significantly low, indicating the weaknesses in the implementation of the technology. The lower values of the sustainability index suggested that there could be shortfalls in the implementation or application of the new crop production technology in a given situation. The analysis brought out that the magnitude of yield realized by crop production technology adopted during the year 2000-01 was sustainable.
14. The agriculture in the State was in the process of transformation from its traditional stage to modernization. The agricultural technology has definitely made contribution for the increase in

the productivity. The technology component alone accounted for to the extent of 60 per cent share in the increased productivity of the selected crops. The magnitude of yield realized by crop production technology adopted during the year 2000-01, was sustainable. However, the full dissemination of the agricultural technology was not observed across the agro-climatic zones. There is need for emphasis on the intensive Research–Extension linkage for exploiting the whole technology on the farmers' field. Similarly, the State has a majority of the cultivable area under rainfed conditions, there is a need for evaluation, management and coordination of new dryland farming technology on priority.

6.3 Policy Implications

The green revolution experienced in India during the last four decades and achievements of self-sufficiency in the foodgrains represented success story for new agricultural technology in the country. The agriculture in Maharashtra has also witnessed a substantial development and change in its production, land use and crop pattern and intensity of input use since the beginning of the plan period. At the same time, there exist wider disparities in the spread of modern technology in agriculture among the different agro-climatic zones because of their excessive dependence on natural factors and constraints in the adoption of advances in crop production technologies. The findings of the present study do have policy implications on the part of State government, other developmental institutions and large number of farmers spread over the length and breadth of the State. The future challenge to reduce the inequalities in income could be met with only through the adoption of new agricultural development strategy of the late 'nineties' and couple of early decades of the this century will have to lay emphasis on evolution of short duration, photo insensitive, high fertilizer responsive, disease resistant, high yielding varieties of different crops, improvement in cultural practices, speedy and effective diffusion of new crop production technologies, increased supplies and use of new forms of inputs, evolution and spread of suitable dryland technology, effective conservation of soil and water, exploitation of additional water resources which have remained yet to be developed, development of infrastructural facilities to support and sustain requirements of increased agricultural production and providing appropriate price incentive to the farm producers.

Some crop specific policy prescription will prove to be of great use in change, *rabi* sorghum and pearlmillet happen to be the major cereal crops of the State, their productivities were quite low for want of varieties with high yield potential. The plant breeders have challenges of evolving suitable varieties of these crops. Though the newly evolved varieties possess high yielding potential and great promises for the future, the same have not been adopted on wider scale. Some of the existing varieties of groundnut, cotton and wheat deserve due consideration for their widespread adoption. The development of high yielding varieties suitable to individual eco-systems, therefore, becomes highly essential in the present context. The research efforts aiming at evolution of low cost technologies of foodgrain crops suitable for rainfed conditions are expected to contribute for increased production in the State.

The existing use levels of plant nutrients need to be increased so as to close down the gap between optimal and existing use levels of plant nutrients for most of the crops in all the agro-climatic zones. There was no balanced use of fertilizers and the farmers used mostly nitrogenous fertilizers. There was negligible use of insecticides and pesticides. The efforts to foster increased use of plant nutrients and plant protection measures would help the growers to increase farm productivity. Sugarcane is the important cash crop, however, its productivity is low and stagnant compared to its potential yield. Immediate efforts are therefore, needed for evolving a breakthrough in the production technology of this crop and to optimize use of different inputs to increase the productivity.

Maharashtra possesses tremendous potentials for development of its agriculture. A strong determination on the part of policy makers, research scientists, extension workers and farmers to plan and execute the various agricultural development programmes more efficiently and effectively would certainly result into bringing a high level prosperity to the farmers of Maharashtra.

Appendices

Appendix-I

Details of area coverage of individual region in Maharashtra

Region	*Sr. No*	*District*
Konkan	1	Mumbai
	2	Thane
	3	Raigad
	4	Ratnagiri
	5	Sindhudurg
Western Maharashtra	1	Kolhapur
	2	Sangli
	3	Satara
	4	Pune
	5	Solapur
	6	Ahmednagar
	7	Nasik
	8	Dhule
	9	Jalgaon
	10	Nandurbar
Marathwada	1	Aurangabad

(Contd.)

(Contd.)

Region	*Sr. No*	*District*
	2	Jalana
	3	Parbhani
	4	Hingoli
	5	Beed
	6	Nanded
	7	Latur
	8	Osmanabad
Vidarbha	1	Buldhana
	2	Akola
	3	Amrawati
	4	Yaotmal
	5	Washim
	6	Vardha
	7	Nagpur
	8	Bhandara
	9	Chandrapur
	10	Gadchiroli
	11	Gondia

Appendix-II

Details of agro-climatic zones of Maharashtra

Zone	*Title of the zone*	*Rainfall (mm)*	*Soil types*	*Major natural vegetation*	*Crops grown*
I	Very high rainfall with lateritic soils	2500 to 3000	Laterites and lateritic soils.	Moist deciduous forest and tree savanna.	Paddy, wal, nagali and hill millets like wari, sawa.
II	Very high rainfall with non-lateritic soils	2250 to 3000	Non-lateritic loam of brown to red colour formed from trap rock.	Moist deciduous open forest with shrub tree savanna.	Paddy, wal, nagali, and hill millets and *rabi* crops like wheat, gram and fruit crops like banana, chiku, areca palm and coconut palm.
III	Ghat zone	3000 to 5000	Mostly forest land and *warkas* soils.	Deciduous to semi-deciduous forest.	Inferior millets like wari, sawa, niger.
IV	Transition zone - A	1750 to 2500	Mainly red to brown with varying depth and texture.	Tree and shrub savanna.	Paddy, nagali and niger.
V	Transition zone - B	700 to 1750	Predominantly grayish black.	Shrub and low scattered savanna.	*Kharif* crops like bajra, gr.nut, cotton, jowar and paddy.

(Contd.)

(Contd.)

Zone	*Title of the zone*	*Rainfall (mm)*	*Soil types*	*Major natural vegetation*	*Crops grown*
VI	Scarcity zone	500 to 700	Medium black calcareous soils formed from trap with varying depth and texture.	Low scattered shrubs with discontinuous thorny thickets.	*Kharif* (excluding paddy) crops cum *rabi* crops mainly depending upon soil depth and texture class.
VII	Assured rainfall zone with mainly *kharif* cropping	700 to 900	Medium black, calcareous soils formed from trap with varying depth and texture.	Low scattered shrubs and pseudostoopes.	Mainly *kharif* crops but generally excluding paddy.
VIII	Moderate to moderately high rainfall with soils formed from trap	900 to 1250	Brown to dark calcareous soils formed from trap with varying depth.	Low scattered shrub savanna with discontinuous thickets.	*Kharif* crops of all types and *rabi* crops like wheat and gram on heavier and deeper soils.
IX	High rainfall zone with soils formed from mixed rocks	1250 to 1700	Yellowish brown to red soils of mixed parentage.	Deciduous open forest with thickets.	*Kharif* crops with paddy as the predominant and jowar, wheat, seasamum, linseed as *rabi* crops.

Appendix-III

Coverage of agro-climatic zones of the Maharashtra

Zone	Tahsils and districts
I	Whole Ratnagiri and Sindhudurg districts and western border strip of Chandgad, Ajra, Gargoti, Radhanagari, Bavada and Shahuwadi tahsils of Kolhapur district.
II	Whole Thana, Raigad and greater Mumbai districts and parts of Igatpuri, Trimbak, Peint and Surgana tahsils of Nasik district.
III	Hilly lying terrain round about the Ghats of Amboli, Fonda and Amba in Konkan and part of Kolhapur district; Koyana and Mahabaleshwar in Satara district; Lonawala and Khandala in Pune district; Kaloobai hill in Akole tahsil of Ahmednagar district; and Igatpuri and Trimbak of Nasik district.
IV	Western hilly flank of Kolhapur; western part of Shirala tahsil of Sangli district; central strip of Patan and Jaoli tahsils of Satara district; western part of Bhor, Velhe and Paud tahsils, central part of Vadgaon tahsil and western strip of Khed, Ambegaon and Junnar tahsils of Pune district; western side of Akole tahsil of Ahmednagar district; Eastern part of Trimbak, Peint and whole of western half of Surgana tahsil of Nasik district.
V	Eastern strip of Kolhapur district; Shirala tahsil of Sangli district; central strip of Satara district; central part of Bhor and eastern part of Paud, Junnar, Ambegaon and Haveli of Pune district; central portion of Akole

(Contd.)

(Contd.)

Zone	*Tahsils and districts*
V	Eastern strip of Kolhapur district; Shirala tahsil of Sangli district; central strip of Satara district; central part of Bhor and eastern part of Paud, Junnar, Ambegaon and Haveli of Pune district; central portion of Akole tahsil of Ahmednagar district; western of boundary of Sinnar tahsil, eastern part of Igatpuri tahsil, central portion of Nasik tahsil, eastern portion of Satana tahsil in Nasik district; western part covering Akkalkuwa and Akrani tahsils of and western part of Taloda, Nandurbar, Sakri tahsils in Dhule district.
VI	Shirol tahsil in Kolhapur district; entire Sangli, Satara, Pune, Ahmednagar, Nasik, Dhule districts (excluding the portions already mentioned above); western strip of Amalner to Chalisgaon in Jalgaon district; western strip of Beed district; Paranda and Bhoom tahsils of Osmanabad district; whole part of Solapur district excluding eastern strip of Barshi, south Solapur tahsil and whole Vaijapur, Gangapur and Paithan tahsils of Aurangabad district.
VII	Remaining parts of Solapur, Osmanabad, Latur, Beed, Aurangabad, Jalna and Jalgaon districts; Shirpur tahsil of Dhule district; Buldhana and Akola districts excluding the southern part of Washim and Mangrulpir tahsils; Amrawati district excluding eastern Chandur and Morshi tahsils; western half of Parabhani district and southern Nanded district covering Kandhar, Mukhed, Billoli, and Degloor tahsils.
VIII	Remaining part of Nanded, Parbhani, Akola and Amrawati districts; whole of Yeotmal and Wardha districts; whole of Nagpur excluding the eastern strip covering parts in Umarkhed, Nagpur and Ramtek tahsils and western part of Chandrapur
IX	The whole Bhandar district and remaining parts of Nagpur, and Chandrapur and Gadchiroli districts are included in this zone. It forms extreme eastern part of the State.

Appendix-IV

Cropping pattern in different agro-climatic zones of Maharashtra during the year 1980-81

(Area in lakh hactares)

Crop	*Agro-climatic zones*									*Maharashtra*
	I	*II*	*III*	*IV*	*V*	*VI*	*VII*	*VIII*	*IX*	
Paddy	1.5	2.89	0.53	1.32	0.77	0.7	1.21	0.4	5.26	14.58
Kharif sorghum	--	--	0.01	0.75	1.18	6.98	10.97	8.63	1.19	29.71
Pearlmillet	--	--	0.08	0.16	1.06	10.56	2.97	0.3	0.21	15.34
Rabi sorghum	--	--	0.09	0.77	1.7	21.07	8.72	0.82	1.81	34.98
Wheat	--	0.04	0.07	0.46	0.55	3.76	3.75	1.31	0.69	10.63
Other cereals	0.7	0.19	0.32	0.03	0.34	0.76	1.18	0.78	0.22	4.52
Total cereals	2.2	3.12	1.1	3.49	5.6	43.83	28.8	12.24	9.38	109.76
Tur	--	0.04	0.01	--	0.01	2.1	3.12	0.94	0.22	6.44

(Contd.)

(Contd.)

Crop	Agro-climatic zones									Maharashtra
	I	II	III	IV	V	VI	VII	VIII	IX	
Gram	--	0.01	0.03	0.17	0.25	1.49	1.6	0.3	0.24	4.09
Total pulses	0.08	0.18	0.13	0.46	1.18	9.91	10.58	2.68	1.95	27.15
Groundnut	--	--	0.06	0.68	0.92	2.79	2.05	0.38	0.07	6.95
Cotton	--	--	--	--	0.14	4.31	13.65	7.22	0.18	25.5
Sugarcane	--	--	--	0.42	0.36	1.77	0.55	0.08	0.01	3.19
Fruits & vegetables	0.24	0.08	0.02	0.24	0.22	0.93	0.77	0.18	0.08	2.76
Gross cropped area	3.57	5.03	2.07	7.9	9.77	70.37	58.68	24.93	14.1	196.42

Source : Season and crop report, 1982-83, Commissioner of Agriculture, Maharashtra

Appendix-V

Cropping pattern in different agro-climatic zones of Maharashtra during the year 2000-01

(Area in lakh hactares)

Crop	*Agro-climatic zones*									*Maharashtra*
	I	*II*	*III*	*IV*	*V*	*VI*	*VII*	*VIII*	*IX*	
Paddy	1.67	2.76	0.25	0.88	0.73	1.28	0.61	1.37	5.55	15.12
Kharif sorghum	0.01	0.00	0.09	0.11	0.69	3.26	9.26	5.51	0.17	19.10
Pearlmillet	0.00	0.00	0.31	1.50	1.76	10.80	3.51	0.12	0.00	18.00
Rabi sorghum	0.01	0.00	0.16	1.84	2.24	18.71	7.83	0.70	0.36	31.84
Wheat	0.01	0.001	0.07	0.34	0.44	2.90	2.08	1.35	0.33	7.52
Other cereals	0.36	0.46	0.15	0.58	0.77	2.94	1.26	0.06	0.07	6.65
Total cereals	2.06	3.22	1.03	5.24	6.63	39.88	24.56	9.11	6.48	98.23

(Contd.)

(Contd.)

Crop	Agro-climatic zones									Maharashtra
	I	II	III	IV	V	VI	VII	VIII	IX	
Tur	0.01	0.04	0.02	0.05	0.27	2.13	4.79	3.32	0.33	10.96
Gram	0.003	0.01	0.01	0.06	0.22	1.46	3.82	1.53	0.02	7.14
Total pulses	0.16	0.32	0.18	0.77	1.62	8.37	15.16	7.89	1.10	35.57
Groundnut	0.11	0.01	0.20	0.63	0.65	2.11	0.89	0.30	0.02	4.90
Cotton	0.0001	0.00	0.01	0.02	0.29	3.61	15.51	10.91	0.42	30.77
Sugarcane	0.07	0.00	0.20	0.69	0.57	2.71	1.29	0.39	0.02	5.95
Fruits & vegetables	2.15	1.18	0.19	0.83	0.79	4.44	3.88	1.63	0.52	15.62
Gross cropped area	4.69	4.78	2.00	8.85	11.16	65.14	68.79	37.31	9.96	212.69

Source : Season and crop report, 1982-83, Commissioner of Agriculture, Maharashtra

Appendix-VI

Per hectare recommended level of the crucial inputs for crops in the State

Sr. No.	Crop	1980-81					2000-01				
		Seed (kg.)	*FYM (Qtl.)*	*Fertilizer doses (Kg)*			*Seed (kg.)*	*FYM (Qtl.)*	*Fertilizer doses (Kg)*		
				N	*P*	*K*			*N*	*P*	*K*
1	*Kharif* Sorghum (Rainfed)	10	100	40	40	--	10	100	80	40	40
2	*Kharif* paddy (Rainfed)	60-70 (Dibling) 80-100 (Sowing)	75	80	40	--	40-45 (seedling) 80-100 (sowing)	75	100	50	--
3	Pearlmillet (Rainfed)	5 (Local)	50	25	25	--	2.5-3 (HB /HYV)	50	60	60	30
4	Gr.nut (Rainfed)	110-120	62.5	20	40	30	80-100	62.5	20	40	--

(Contd.)

(Contd.)

Sr. No.	Crop	1980-81					2000-01				
		Seed (kg.)	*FYM (Qtl.)*	*Fertilizer doses (Kg)*			*Seed (kg.)*	*FYM (Qtl.)*	*Fertilizer doses (Kg)*		
				N	*P*	*K*			*N*	*P*	*K*
5	Cotton (Rainfed)	8-10 (Local) 1.5-3 (HB)	200	50	50	25	12-15 (Local), 3-4 (HB/HYV)	200	80	40	40
6	Rabi sorghum (Rainfed)	10	55	40	20	--	10	55	60	25	--
7	Wheat (Irrigated)	100	200	120	50	50	100-125	200	80	40	30
8	Sugarcane *(Suru)*	65-70 Qtl.	200	250	115	115	65-70 Qtl.	200	250	115	115
9	Sugarcane *(Ratoon)*	--	200	250	115	115	--	200	250	115	115

Source : 1. Krishi Sanchyika Directorate of Agriculture, Maharashtra state, Pune Booklet No 953, September, 1983, pp. 9-17.

2. 'Krishi darshani- 2000-01' published by SAU's Maharashtra

Appendix-VII

Per hectare recommended level of the crucial inputs for crops in the State

Crop / Zones	Number	Low level (C I below 30)			Medium level (C I 31 to 60)			High level (C I above 60)			Average composite index		
	Avg T.I.	1980-81	2000-01	% change	1980-81	2000-01	% change	1980-81	2000-01	% change	1980-81	2000-01	% change
Kharif Sorghum													
VII	Number	11	4		10	15		1	3		22	22	
	Avg T.I.	18.95	21.05	11.08	34.21	39.94	16.75	61.93	77.26	24.75	27.84	41.59	49.41
VIII	Number	20	4		1	17					21	21	
	Avg T.I.	15.54	26.32	69.37	31.57	38.70	22.58	--	--	--	16.30	36.34	122.91
Overall	Number	31	8		11	32		1	3		43	43	
	Avg T.I.	16.75	23.69	41.40	33.97	39.28	15.64	61.93	77.26	24.75	22.21	39.03	75.76
Kharif paddy													
II	Number	19	5		1	15					20	20	
	Avg T.I.	13.74	25.26	83.84	31.06	40.63	30.81	--	--	--	14.61	36.79	151.87
V	Number	19	8		1	12					20	20	

(Contd.)

(Contd.)

Crop / Zones	Number	Low level (C I below 30)			Medium level (C I 31 to 60)			High level (C I above 60)			Average composite index		
	Avg T.I.	1980-81	2000-01	% change	1980-81	2000-01	% change	1980-81	2000-01	% change	1980-81	2000-01	% change
Kharif	Avg T.I.	18.28	26.29	43.82	32.64	41.02	25.67	--	--	--	19.00	35.13	84.90
IX	Number	20	4			16					20	20	
	Avg T.I.	18.42	26.32	42.89		42.11	--	--	--	--	18.42	38.95	111.47
Overall	Number	58	17		2	43					60	60	
	Avg T.I.	16.84	25.99	54.35	31.85	41.29	29.64	--	--	--	17.34	36.96	113.11
Pearlmillet													
V	Number	19	14		1	6					20	20	
	Avg T.I.	8.03	19.35	140.97	31.58	37.50	18.75	--	--	--	9.21	24.80	169.29
VI	Number	20	16			4					20	20	
	Avg T.I.	9.21	20.83	126.17		39.58	--	--	--	--	9.21	24.58	166.88
VIII	Number	14	13			1					14	14	
	Avg T.I.	9.02	23.40	159.42		33.33	--	--	--	--	9.02	24.11	167.29

(Contd.)

(Contd.)

Crop / Zones	Number / Avg T.I.	Low level (C I below 30)			Medium level (C I 31 to 60)			High level (C I above 60)			Average composite index		
		1980-81	2000-01	% change	1980-81	2000-01	% change	1980-81	2000-01	% change	1980-81	2000-01	% change
Kharif													
Overall	Number	53	43		1	11					54	54	
	Avg T.I.	8.74	21.13	141.79	31.58	37.88	19.94	--	--	--	9.16	24.54	167.88
Groundnut													
V	Number	5	4		14	12		2	3		21	19	
	Avg T.I.	20.00	20.61	3.05	31.08	41.23	32.66	69.06	78.95	14.32	32.06	42.84	33.64
VII	Number	11	4		9	15		1	1		21	20	
	Avg T.I.	15.79	17.05	7.98	32.56	47.33	45.36	61.35	68.42	11.52	25.15	42.33	68.33
Overall	Number	16	8		23	27		3	4		42	39	
	Avg T.I.	17.11	18.83	10.08	31.66	44.62	40.94	66.49	76.32	14.78	28.60	42.58	48.87
Cotton													
VII	Number	5	3		14	16					19	19	
	Avg T.I.	24.21	28.12	16.15	37.59	41.78	11.15	--	--	--	34.07	39.62	16.30

(Contd.)

(Contd.)

Crop / Zones	Number	Low level (C I below 30)			Medium level (C I 31 to 60)			High level (C I above 60)			Average composite index		
	Avg T.I.	1980-81	2000-01	% change	1980-81	2000-01	% change	1980-81	2000-01	% change	1980-81	2000-01	% change
VIII	Number	11	2		9	17			1		20	20	
	Avg T.I.	19.62	23.99	22.27	41.69	45.82	9.91	--	73.68	--	29.55	45.03	52.38
Overall	Number	16	5		23	33			1		39	39	
	Avg T.I.	21.05	26.47	25.71	39.19	43.86	11.91	--	73.68	--	31.75	42.40	33.52
Rabi sorghum													
VI	Number	16	11		4	9					20	20	
	Avg T.I.	16.78	18.66	11.20	34.47	39.77	15.38	--	--	--	20.32	28.16	38.59
VII	Number	15	13		1	3					16	16	
	Avg T.I.	16.78	17.19	2.44	33.64	37.37	11.09	--	--	--	17.83	20.97	17.61
Overall	Number	31	24		5	12					36	36	
	Avg T.I.	16.78	17.86	6.46	34.30	39.17	14.18	--	--	--	19.21	24.97	29.94
Wheat													
VI	Number	15	9		3	9					18	18	

(Contd.)

(Contd.)

Crop / Zones	Number	Low level (C I below 30)			Medium level (C I 31 to 60)			High level (C I above 60)			Average composite index		
	Avg T.I.	1980-81	2000-01	% change	1980-81	2000-01	% change	1980-81	2000-01	% change	1980-81	2000-01	% change
Kharif													
	Avg T.I.	16.14	21.64	34.08	36.84	42.69	15.88	--	--	--	19.59	32.17	64.19
VII	Number	17	10		1	8					18	18	
	Avg T.I.	17.84	22.53	26.29	31.98	43.79	36.93	--	--	--	18.63	31.98	71.69
VIII	Number	18			1	19					19	19	
	Avg T.I.	18.24			32.83	47.12	43.53	--	--	--	19.01	47.12	147.90
Overall	Number	50	19		5	36					55	55	
	Avg T.I.	17.47	22.11	26.52	35.07	45.27	29.11	--	--	--	19.07	37.27	95.41
Sugarcane *(Suru)*													
VI	Number	3	1		4	2		13	17		20	20	
	Avg T.I.	17.98	21.65	20.41	41.11	48.64	18.32	66.51	73.15	9.98	54.15	68.12	25.80
VII	Number	2			6	7		4	5		12	12	
	Avg T.I.	23.49			46.57	55.07	18.25	61.23	69.81	14.01	47.61	61.21	28.57

(Contd.)

(Contd.)

Crop / Zones	Number / Avg T.I.	*Low level (C I below 30)* 1980-81	2000-01	% change	*Medium level (C I 31 to 60)* 1980-81	2000-01	% change	*High level (C I above 60)* 1980-81	2000-01	% change	*Average composite index* 1980-81	2000-01	% change
Kharif													
Overall	Number	5	1		10	9		17	22		32	32	
	Avg T.I.	20.18	21.65	7.26	44.39	53.64	20.85	65.27	72.39	10.91	51.70	65.53	26.76
Sugarcane *(Ratton)*													
VI	Number	9	5		8	7			5		17	17	
	Avg T.I.	19.46	27.15	39.52	33.97	39.33	15.78	-	71.04	100.00	26.29	45.07	71.46
VII	Number	10	7		7	6			4		17	17	
	Avg T.I.	18.53	25.16	35.78	33.64	36.93	9.78	-	72.89	100.00	24.75	40.54	63.81
Overall	Number	19	12		15	13			9		34	34	
	Avg T.I.	18.97	25.99	37.00	33.82	38.22	13.03	-	71.86	100.00	25.52	42.81	67.75

Appendix-VIII(A)

Per hectare itemwise cost of cultivation of *kharif* sorghum

Sr. No.	Cost items	1980-81*			2000-01		
		VII	*VIII*	*Overall*	*VII*	*VIII*	*Overall*
1	**Labour cost**						
	a. human labour	3859.14	2233.78	2675.03	3587.14	3672.69	3635.38
	b. buloock power	3116.26	2023.52	2320.39	2419.41	1777.70	2057.57
	c. Machine power	0.00	30.10	21.96	668.42	1182.37	958.22
	sub-total	6975.40	4287.41	5017.38	6674.97	6632.76	6651.17
2	**Material input**						
	a. Seed	674.50	586.06	610.27	492.20	630.17	570.00
	b. Manure	2487.93	263.37	866.09	1676.15	327.64	915.77
	c. Fertilizers	505.66	32.05	160.36	603.10	943.14	794.84
	d. Plant protection	9.84	7.11	7.85	1.38	0.00	0.60
	Sub total	3677.93	888.60	1644.57	2772.83	1900.95	2281.21
3	**Other costs**						
	a. Irrigation	0.00	0.00	0.00	0.00	0.00	0.00
	b. Incedental	243.03	23.70	83.12	181.73	93.14	131.78
	c. Repairs	53.76	77.91	71.40	162.13	141.57	150.54
	Sub-total	296.79	101.61	154.53	343.86	234.71	282.31
4	**Indirect costs**						
	a. Interst on working capital	614.65	308.97	391.91	576.95	515.66	542.39

(Contd.)

(Contd.)

Sr. No.	*Cost items*	*1980-81**			*2000-01*		
		VII	*VIII*	*Overall*	*VII*	*VIII*	*Overall*
	b. Depreciation on farm impl.	483.64	61.35	175.77	431.89	532.05	488.37
	c. Land revenue and taxes	128.58	11.08	42.91	30.64	32.93	31.93
	d. Rental value of land	1179.35	621.55	772.92	2166.30	1698.42	1902.48
	e. Interst on fixed capital	331.31	37.12	116.83	566.24	674.66	627.37
	Sub- total	2737.53	1040.07	1500.34	3772.02	3453.72	3592.54
5	**Total cost of cultivation**	13687.65	6317.69	8316.81	13563.68	12222.14	12807.23
6	**Gross returns**	7847.56	3795.80	4894.99	13181.66	10388.11	11606.47
7	**Net returns**	-5840.09	-2521.89	-3421.82	-382.02	-1834.03	-1200.76
8	**Per quintal cost**	894.62	888.56	891.25	591.01	721.07	654.54

Appendix-VIII(B)

Per hactare itemwise cost of cultivation of *Kharif* paddy

Sr. No.	Cost items	*1980-81**				*2000-01*			
		II	*V*	*IX*	*Overall*	*II*	*V*	*IX*	*Overall*
1	**Labour cost**								
	a. human labour	5108.19	9022.56	5052.75	5437.46	1819.04	2968.57	5127.25	2532.86
	b. buloock power	2162.44	2028.14	1837.85	1977.40	1541.82	1371.84	2165.51	1588.02
	c. Machine power	0.00	60.47	0.00	5.54	512.06	1159.34	0.00	590.16
	sub-total	7270.64	11111.18	6890.60	7420.40	3872.92	5499.75	7292.76	4711.04
2	**Material input**								
	a. Seed	772.24	664.80	535.10	636.19	1536.46	1331.87	1228.73	1447.74
	b. Manure	127.84	1654.10	488.24	459.52	1413.40	208.79	331.49	990.58
	c. Fertilizers	431.87	821.92	34.81	310.88	218.24	584.03	1168.29	431.39
	d. Plant protection	5.47	0.00	0.00	2.06	0.00	0.00	0.00	0.00
	Sub total	1337.42	3140.82	1058.14	1408.66	3168.10	2124.69	2728.51	2869.71
3	**Other costs**								
	a. Irrigation	0.00	0.00	0.00	0.00	0.00	0.00	0.00	0.00

(Contd.)

(Contd.)

Sr. No.	Cost items	1980-81*				2000-01			
		II	V	IX	Overall	II	V	IX	Overall
	Sub total	1337.42	3140.82	1058.14	1408.66	3168.10	2124.69	2728.51	2869.71
3	**Other costs**								
	a. Irrigation	0.00	0.00	0.00	0.00	0.00	0.00	0.00	0.00
	b. Incedental	146.72	421.36	39.49	302.26	116.82	173.13	334.50	159.37
	c. Repairs	84.44	48.95	93.14	255.70	181.78	345.86	405.67	249.81
	Sub-total	231.16	470.31	132.63	557.96	298.60	518.99	740.17	409.18
4	**Indirect costs**								
	a. Interst on working capital	452.80	697.63	399.23	475.56	430.22	451.47	598.77	458.06
	b. Depreciation	184.49	1108.89	165.48	1073.60	314.42	994.20	1337.53	609.28
	c. Land revenue and taxes	7.86	263.56	19.82	156.03	14.62	21.84	27.78	18.06
	d. Rental value of land	1313.27	1342.72	2423.34	1788.34	2801.51	4020.29	4229.25	3274.73

(Contd.)

(Contd.)

Sr. No.	Cost items	*1980-81**				*2000-01*			
		II	*V*	*IX*	*Overall*	*II*	*V*	*IX*	*Overall*
	e. Interst on fixed capital	242.12	211.13	163.69	197.54	520.23	1891.46	2650.49	1124.08
	Sub-total	2200.54	3623.93	3171.56	3691.07	4081.00	7379.26	8843.82	5484.22
5	**Total cost of cultivation**	11039.75	18346.25	11252.93	13078.08	11420.62	15522.69	19605.26	13474.15
6	**Gros. returns**	7926.77	9637.69	14658.95	11666.19	16896.80	24252.75	25542.19	19756.76
7	**Net returns**	0.00	-8708.56	3406.02	-1411.89	5476.18	8730.06	5936.93	6282.61
8	**Per quintal cost**	1113.05	1698.73	831.09	1113.36	561.49	520.37	832.50	587.20

(- Inflated periods)*

Appendix-VIII(C)

Per hactare itemwise cost of cultivation of pearlmillet

Sr. No.	Cost items	1980-81*				2000-01			
		V	*VI*	*VII*	*Overall*	*V*	*VI*	*VII*	*Overall*
1	**Labour cost**								
	a. human labour	1357.98	2027.08	1979.74	1745.09	3296.57	2057.98	2447.54	2476.89
	b. buloock power	2204.09	2002.22	2209.58	2120.49	2810.45	1694.22	1324.01	1890.85
	c. Machine power	0.00	0.42	0.00	0.18	316.45	236.57	180.23	243.16
	sub-total	3562.07	4029.72	4189.32	3865.75	6423.47	3988.77	3951.78	4610.89
2	**Material input**								
	a. Seed	304.75	238.57	257.71	268.95	328.65	238.57	289.01	274.58
	b. Manure	14.49	49.81	178.35	57.45	198.77	335.99	0.00	216.14
	c. Fertilizers	2.61	20.47	39.69	16.47	496.56	244.18	593.90	397.34
	d. Plant protection	18.79	0.00	0.00	7.69	1.75	0.23	0.29	0.64
	Sub total	340.65	308.85	475.75	350.55	1025.73	818.97	883.20	888.70
3	**Other costs**								
	a. Irrigation	0.00	0.00	0.00	0.00	0.00	0.00	0.00	0.00

(Contd.)

(Contd.)

Sr. No.	Cost items	1980-81*				2000-01			
		V	VI	VII	Overall	V	VI	VII	Overall
	b. Incedental	40.56	97.08	216.93	94.55	101.49	124.42	76.88	106.55
	c. Repairs	32.84	58.27	40.35	44.78	130.93	79.64	114.21	101.61
	Sub-total	73.40	155.35	257.28	139.33	232.42	204.06	191.09	208.16
4	**Indirect costs**								
	a. Interst on working capital	883.07	983.43	1033.34	950.93	410.07	292.50	294.59	323.51
	b. Depreciation on farm impl.	90.40	104.63	141.70	105.18	806.28	216.84	531.71	448.67
	c. Land revenue and taxes	6.84	9.29	12.57	8.85	26.47	11.62	27.60	19.48
	d. Rental value of land	383.59	504.60	677.78	484.84	1934.24	1039.22	1006.19	1263.04
	e. Interst on fixed capital	74.19	111.14	149.37	102.59	851.88	244.90	600.77	491.56
	Sub- total	1438.10	1713.09	2014.77	1652.40	4028.93	1805.08	2460.86	2546.27
5	**Total cost of cultivation**	5414.21	6207.01	6937.11	6008.04	11710.55	6816.88	7486.93	8254.01
6	**Gross returns**	2301.56	3027.59	4066.69	2909.06	11764.23	6305.05	6202.76	7695.13
7	**Net returns**	-3112.65	-3179.42	-2870.42	-3098.97	53.68	-511.83	-1284.17	-558.88
8	**Per quintal cost**	2082.39	1474.35	1077.19	1655.11	731.91	720.60	821.84	745.74

(- Inflated periods)*

Appendix-VIII(D)

Per hactare itemwise cost of cultivation of Groundnut

Sr. No.	Cost items	1980-81*			2000-01		
		V	VII	Overall	V	VII	Overall
1	**Labour cost**						
	a. human labour	5188.56	3978.57	4495.74	6285.47	3445.32	4559.07
	b. buloock power	3725.12	2301.43	2909.93	3522.24	2324.52	2794.20
	c. Machine power	716.64	0.00	306.30	1128.28	656.41	841.45
	sub-total	9630.31	6279.99	7711.97	10935.99	6426.25	8194.72
2	**Material input**						
	a. Seed	3702.04	3338.87	3494.09	3458.66	2861.44	3095.64
	b. Manure	2313.80	479.25	1263.37	3344.99	605.13	1679.55
	c. Fertilizers	340.56	493.81	428.31	558.82	679.89	632.41
	d. Plant protection	0.00	0.00	0.00	27.96	0.96	11.55
	Sub total	6356.40	4311.93	5185.77	7390.43	4147.42	5419.15
3	**Other costs**						
	a. Irrigation	0.00	0.00	0.00	0.00	0.00	0.00

(Contd.)

(Contd.)

Sr. No.	Cost items	1980-81*			2000-01		
		V	VII	Overall	V	VII	Overall
	b. Incedental	372.98	205.11	276.86	179.73	136.13	153.23
	c. Repairs	87.75	80.99	83.88	295.69	98.78	176.00
	Sub-total	460.73	286.10	360.74	475.42	234.91	329.22
4	**Indirect costs**						
	a. Interst on working capital	4180.21	2699.22	3332.22	1129.24	661.55	844.95
	b. Depreciation on farm impl.	421.29	431.04	426.87	1060.07	538.70	743.15
	c. Land revenue and taxes	266.28	146.93	197.94	48.62	19.30	30.80
	d. Rental value of land	3073.32	1527.54	2188.23	4312.91	2066.47	2947.40
	e. Interst on fixed capital	765.04	444.48	581.49	1300.42	480.14	801.81
	Sub- total	8706.14	5249.21	6726.76	7851.25	3766.16	5368.11
5	**Total cost of cultivation**	25153.59	16127.23	19985.24	26653.09	14574.74	19311.20
6	**Gross returns**	18439.90	9165.24	13129.37	26169.16	12514.62	17869.17
7	**Net returns**	-6713.69	-6962.00	-6855.87	-483.93	-2060.12	-1442.03
8	**Per quintal cost**	1896.95	2849.33	2243.01	1426.83	1821.84	1584.43

(- Inflated periods)*

Appendix-VIII(E)

Per hactare itemwise cost of cultivation of Cotton

Sr. No.	Cost items	1980-81*			2000-01		
		VII	VIII	Overall	VII	VIII	Overall
1	**Labour cost**						
	a. human labour	1876.37	2148.39	1989.06	3902.37	3021.94	3348.98
	b. buloock power	1848.04	2450.49	2097.61	2112.04	2308.99	2235.83
	c. Machine power	162.18	273.89	208.46	254.04	250.36	251.73
	sub-total	3886.59	4872.77	4295.12	6268.45	5581.29	5836.54
2	**Material input**						
	a. Seed	1463.41	532.60	1077.81	1010.78	1615.54	1390.90
	b. Manure	187.85	652.83	380.47	1040.75	875.18	936.68
	c. Fertilizers	213.16	207.68	210.89	1181.95	1178.23	1179.61
	d. Plant protection	13.08	65.20	34.67	396.24	102.00	211.30
	Sub total	1877.50	1458.30	1703.84	3629.72	3770.95	3718.49
3	**Other costs**						
	a. Irrigation	0.00	0.00	0.00	0.00	0.00	0.00

(Contd.)

(Contd.)

Sr. No.	Cost items	1980-81*			2000-01		
		VII	VIII	Overall	VII	VIII	Overall
	b. Incedental	34.42	82.32	54.26	76.40	58.50	65.15
	c. Repairs	46.28	37.20	42.52	98.85	78.82	86.26
	Sub-total	80.70	119.52	96.78	175.25	137.32	151.41
4	**Indirect costs**						
	a. Interst on working capital	2150.82	2347.44	2232.27	910.27	882.46	892.79
	b. Depreciation on farm impl.	58.43	75.10	65.34	417.31	499.94	469.25
	c. Land revenue and taxes	13.52	9.50	11.85	44.79	27.15	33.70
	d. Rental value of land	1236.56	1245.05	253.85	2671.35	1852.42	2156.62
	e. Interst on fixed capital	76.89	42.92	62.82	639.63	855.14	775.09
	Sub- total	3536.21	3720.00	2626.13	4683.35	4117.12	4327.45
5	**Total cost of cultivation**	9381.00	10170.59	8721.87	14756.77	13606.68	14033.88
6	**Gross returns**	7419.36	7833.29	7590.84	16296.86	11277.44	13141.93
7	**Net returns**	-1961.64	-2337.30	-1131.03	1540.09	-2329.24	-891.96
8	**Per quintal cost**	2727.03	2801.81	2477.80	1951.95	2482.97	2244.48

(- Inflated periods)*

Appendix-VIII(F)

Per hactare itemwise cost of cultivation of *rabi* sorghum

Sr. No.	Cost items	*1980-81**			*2000-01*		
		VI	*VII*	*Overall*	*VI*	*VII*	*Overall*
1	**Labour cost**						
	a. human labour	1080.72	1096.82	1086.95	3212.96	3384.78	3306.53
	b. buloock power	1263.84	1275.66	1268.42	928.17	1203.69	1078.22
	c. Machine power	36.42	11.87	26.92	696.05	131.04	388.35
	sub-total	2380.98	2384.36	2382.29	4837.18	4719.51	4773.10
2	**Material input**						
	a. Seed	106.22	201.00	142.90	109.79	231.21	175.91
	b. Manure	57.04	169.15	100.43	522.39	82.38	282.77
	c. Fertilizers	29.72	0.00	18.22	316.05	56.47	174.69
	d. Plant protection	0.00	0.00	0.00	8.66	3.27	5.72
	Sub total	192.98	370.15	261.55	956.89	373.33	639.09
3	**Other costs**						
	a. Irrigation	0.00	0.00	0.00	0.00	0.00	0.00

(Contd.)

(Contd.)

Sr. No.	Cost items	1980-81*			2000-01		
		VII	VIII	Overall	VII	VIII	Overall
	b. Incedental	34.42	82.32	54.26	76.40	58.50	65.15
	c. Repairs	46.28	37.20	42.52	98.85	78.82	86.26
	Sub-total	80.70	119.52	96.78	175.25	137.32	151.41
4	**Indirect costs**						
	a. Interst on working capital	2150.82	2347.44	2232.27	910.27	882.46	892.79
	b. Depreciation on farm impl.	58.43	75.10	65.34	417.31	499.94	469.25
	c. Land revenue and taxes	13.52	9.50	11.85	44.79	27.15	33.70
	d. Rental value of land	1236.56	1245.05	253.85	2671.35	1852.42	2156.62
	e. Interst on fixed capital	76.89	42.92	62.82	639.63	855.14	775.09
	Sub- total	3536.21	3720.00	2626.13	4683.35	4117.12	4327.45
5	**Total cost of cultivation**	9381.00	10170.59	8721.87	14756.77	13606.68	14033.88
6	**Gross returns**	7419.36	7833.29	7590.84	16296.86	11277.44	13141.93
7	**Net returns**	-1961.64	-2337.30	-1131.03	1540.09	-2329.24	-891.96
8	**Per quintal cost**	2727.03	2801.81	2477.80	1951.95	2482.97	2244.48

(- Inflated periods)*

Appendix-VIII(G)

Per hactare itemwise cost of cultivation of wheat

Sr. No.	Cost items	1980-81*				2000-01			
		VI	VII	VIII	Overall	VI	VII	VIII	Overall
1	**Labour cost**								
	a. human labour	3050.27	3114.96	3149.93	3101.91	4664.52	4248.96	3589.53	4014.60
	b. buloock power	2308.13	2492.13	1763.49	2236.39	1184.29	865.69	1684.71	1344.43
	c. Machine power	73.77	66.27	0.00	51.35	818.50	906.84	634.31	751.51
	Sub-total	5432.17	5673.35	4913.42	5389.65	6667.31	6021.49	5908.55	6110.54
2	**Material input**								
	a. Seed	1250.73	1187.66	1306.34	1240.71	1241.33	1159.46	1267.32	1231.48
	b. Manure	1408.36	0.00	0.00	485.03	1200.66	2079.44	529.41	1111.39
	c. Fertilizers	1312.06	1361.67	990.09	1246.48	714.87	897.44	1237.61	1025.48
	d. Plant protection	0.00	4.56	0.00	1.79	23.26	24.07	0.00	11.92
	Sub total	3971.14	2553.89	2296.43	2974.01	3180.12	4160.41	3034.34	3380.28
3	**Other costs**								
	a. Irrigation	3088.48	2882.33	2020.83	2725.87	999.87	820.73	785.91	843.70

(Contd.)

(Contd.)

Sr. No.	Cost items	1980-81*				2000-01			
		VI	*VII*	*VIII*	*Overall*	*VI*	*VII*	*VIII*	*Overall*
	b. Incedental	649.67	261.92	403.21	432.76	131.74	131.70	52.74	92.46
	c. Repairs	132.54	122.92	106.91	122.00	129.15	104.60	46.53	81.25
	Sub-total	3870.69	3267.17	2530.96	3280.64	1260.76	1057.03	885.18	1017.41
4	**Indirect costs**								
	a. Interst on working capital	2661.81	2229.59	1808.12	2267.17	622.51	637.68	595.61	613.35
	b. Depreciation on farm impl.	666.01	472.92	503.69	547.54	851.14	745.01	357.90	576.45
	c. Land revenue and taxes	151.74	126.03	81.61	123.16	21.35	31.15	41.51	34.10
	d. Rental value of land	1810.87	2202.15	917.47	1728.21	2738.77	2414.90	1774.29	2169.28
	e. Interst on fixed capital	807.84	470.27	454.84	582.45	1647.60	1774.45	153.42	940.16
	Sub- total	6098.26	5500.97	3765.74	5248.54	5881.36	5603.19	2922.72	4333.33
5	**Total cost of cultivation**	19372.25	16995.38	13506.55	16892.84	16989.55	16842.12	12750.79	14841.56
6	**Gross returns**	10865.20	13212.90	5504.80	10369.29	16560.69	14676.29	10894.77	13220.24
7	**Net returns**	-8507.05	-3782.48	-8001.74	-6523.55	-428.86	-2165.83	-1856.02	-1621.32
8	**Per quintal cost**	1282.93	953.19	1671.60	1179.67	740.29	866.36	770.91	789.97

(- Inflated periods)*

Appendix-VIII(H)

Per hactare itemwise cost of cultivation of *suru* sugarcane

Sr. No.	Cost items	*1980-81**			*2000-01*		
		VI	*VII*	*Overall*	*VI*	*VII*	*Overall*
1	**Labour cost**						
	a. human labour	23125.73	25455.69	23897.89	14623.16	14947.94	14783.36
	b. buloock power	4268.75	4271.45	4269.64	2881.69	2285.35	2587.54
	c. Machine power	2058.85	1818.08	1979.06	3556.43	3908.33	3730.01
	Sub-total	29453.33	31545.22	30146.59	21061.28	21141.62	21100.91
2	**Material input**						
	a. Seed	8061.09	8087.00	8069.68	7288.34	7839.33	7560.12
	b. Manure	5655.71	2899.81	4742.39	2034.22	3092.45	2556.20
	c. Fertilizers	8862.22	4200.00	7317.13	6516.72	5940.59	6232.54
	d. Plant protection	0.00	0.00	0.00	0.00	0.00	0.00
	Sub-total	22579.03	15186.81	20129.20	15839.28	16872.37	16348.86
3	**Other costs**						
	a. Irrigation	3690.93	4918.25	4097.67	3690.93	4918.25	4296.31

(Contd.)

(Contd.)

Sr. No.	Cost items	1980-81*			2000-01		
		VI	VII	Overall	VI	VII	Overall
	b. Incedental	472.21	729.90	557.61	206.19	188.28	197.36
	c. Repairs	582.47	450.05	538.59	367.29	229.25	299.20
	Sub-total	4745.61	6098.20	5193.87	4264.41	5335.78	4792.87
4	**Indirect costs**						
	a. Interst on working capital	6796.78	6708.01	6767.36	2565.63	2717.98	2640.77
	b. Depreciation on farm impl.	1150.33	1013.45	1104.97	914.25	963.60	938.59
	c. Land revenue and taxes	602.91	492.56	566.34	192.89	135.70	164.68
	d. Rental value of land	13594.53	12635.54	13276.72	14318.48	14559.89	14437.56
	e. Interst on fixed capital	1301.51	1301.25	1301.42	1960.62	2619.98	2285.85
	Sub- total	23446.06	22150.81	23016.81	19951.87	20997.15	20467.46
5	**Total cost of cultivation**	80224.03	74981.04	78486.47	61116.84	64346.92	62710.09
6	**Gross returns**	85184.63	78768.62	83088.33	87068.22	88173.53	87613.42
7	**Net returns**	4960.60	3787.58	4601.86	25951.38	23826.61	24903.33
8	**Per quintal cost**	85.04	87.72	85.81	62.25	64.55	63.40

(- Inflated periods)*

Appendix-VIII(I)

Per hactare itemwise cost of cultivation of *ratoon* sugarcane

Sr. No.	Cost items	*1980-81**			*2000-01*		
		VI	*VII*	*Overall*	*VI*	*VII*	*Overall*
1	**Labour cost**						
	a. human labour	12766.68	16075.71	14246.03	13378.53	11508.20	12080.89
	b. buloock power	771.21	1082.94	910.57	571.81	1306.55	1081.57
	c. Machine power	1701.84	958.47	1369.50	119.72	62.16	79.78
	Sub-total	15239.73	18117.12	16526.11	14070.06	12876.91	13242.25
2	**Material input**						
	a. Seed	0.00	0.00	0.00	0.00	0.00	0.00
	b. Manure	1829.58	1570.34	1713.68	440.14	1398.60	1105.12
	c. Fertilizers	3913.23	2510.60	3286.16	5752.47	3997.54	4534.90
	d. Plant protection	0.00	0.00	0.00	0.00	0.00	0.00
	Sub-total	5742.81	4080.94	4999.85	6192.61	5396.14	5640.02
3	**Other costs**						
	a. Irrigation	4249.04	4005.72	4140.26	4249.04	4005.72	4080.22

(Contd.)

(Contd.)

Sr. No.	Cost items	1980-81*			2000-01		
		VI	VII	Overall	VI	VII	Overall
	b. Incedental	525.97	410.86	474.51	161.24	254.74	226.11
	c. Repairs	433.50	341.42	392.33	152.92	345.32	286.41
	Sub-total	5208.51	4758.00	5007.10	4563.20	4605.78	4592.74
4	**Indirect costs**						
	a. Interst on working capital	3096.18	3127.71	3110.28	1458.34	1232.96	1301.97
	b. Depreciation on farm impl.	1142.30	665.23	929.02	668.40	1880.00	1509.01
	c. Land revenue and taxes	539.92	464.79	506.33	238.97	149.58	176.95
	d. Rental value of land	10888.34	10552.87	10738.36	11546.65	11420.43	11459.08
	e. Interst on fixed capital	1308.66	1276.13	1294.12	1439.55	3589.46	2931.16
	Sub- total	16975.40	16086.73	16578.11	15351.91	18272.43	17378.17
5	**Total cost of cultivation**	43166.45	43042.79	43111.17	40177.78	41151.26	40853.18
6	**Gross returns**	68569.57	66105.97	67468.18	70713.69	69420.08	69816.18
7	**Net returns**	25403.12	23063.17	24357.01	30535.91	28268.82	28963.00
8	**Per quintal cost**	56.15	59.87	57.75	50.52	52.68	52.01

(*- Inflated periods)

Appendix-IX(A)

Results of Cobb-Douglas production function of *kharif* sorghum in different agro-climatic zones

Particulars	*1980-81*			*2000-01*		
	VII	*VIII*	*Overall*	*VII*	*VIII*	*Overall*
Constant (a)	-1.1210 (0.7513)	-0.7004 (0.7620)	1.1789 (0.3139)	0.1669 (0.8218)	-0.1868 (1.0008)	1.2926 (0.6284)
Human labour in Mandays (X1)	0.9254** (0.3838)	0.8399 (0.4297)	0.5702*** (0.1104)	0.2444 (0.3080)	0.4172 (0.3305)	0.3706 (0.3206)
Bullock power in pairdays (X2)	0.0397 (0.2983)	0.4274* (0.2730)	0.4916 (0.2011)	0.0243 (0.0912)	0.0709 (0.1013)	0.2681 (0.1945)
Seed in Kg. (X3)	0.1223 (0.4684)	-0.2460 (0.6257)	-0.1549 (0.3706)	0.1608 (0.5666)	-0.2241 (0.4482)	0.2784** (0.1130)
Manures and fertlizers in Rs. (X4)	0.1669*** (0.0572)	0.0311 (0.0419)	0.0711** (0.0409)	0.1388** (0.0556)	-0.0619 (0.0936)	0.2171** (0.0524)
Other working capital in Rs. (X5)	-0.0598 (0.0764)	-0.1821** (0.0942)	0.0486 (0.0284)	0.0522 (0.1312)	0.2766** (0.1022)	0.0517 (0.0796)
R2	0.7749	0.7784	0.7900	0.5723	0.5214	0.4348
F value	4.81	4.61	27.84***	4.28***	3.26***	5.69***

(Figures in parentheses indicate the standard errors of the respective coefficients.)

*' ** and *** indicates the level of significance at 10.5 and 1 per cent. NS= non-significant.

Appendix-IX(B)

Results of Cobb-Douglas production function of *kharif* paddy in different agro-climatic zones

Particulars	*1980-81*				*2000-01*			
	II	*V*	*IX*	*Overall*	*II*	*V*	*IX*	*Overall*
Constant (a)	-1.7258 (1.7311)	-1.4117 (1.2447)	-0.4818 (1.5555)	-0.8516 (1.0252)	4.6645 (2.1324)	-1.7149 (0.9870)	-2.5686 (1.1764)	0.1605 (0.7726)
Human labour in Mandays (X1)	-0.2409 (0.5892)	-0.1411 (0.2234)	0.1557 (0.4744)	0.3706 (0.2277)	0.2023 (0.2093)	0.3880* (0.2693)	0.4716*** (0.1436)	0.2065** (0.0989)
Bullock power in pairdays (X2)	0.5240 (1.1146)	0.0595 (0.2080)	0.2193 (0.4248)	0.1432 (0.2680)	-0.2674*** (0.0951)	0.0898 (0.1241)	0.1522* (0.0953)	0.0432 (0.0664)
Seed in Kg. (X3)	1.0152 (1.0775)	1.0884** (0.6216)	0.5657 (1.0568)	0.2453** (0.1357)	-1.7804* (1.1284)	0.7458 (0.5883)	0.9577* (0.5867)	0.4242 (0.4258)
Manures and fertlizers in Rs. (X4)	0.4864*** (0.1378)	0.3153** (0.0801)	0.0115 (0.0779)	0.2692*** (0.0643)	-0.0707 (0.1007)	0.0818 (0.1681)	0.0933* (0.0684)	0.1948*** (0.0609)
Other working capital in Rs. (X5)	-0.0473 (0.1349)	0.0475 (0.0936)	0.1074 (0.1365)	0.0905* (0.0658)	0.1043* (0.0662)	0.2278** (0.1006)	0.2373*** (0.0715)	0.0395*** (0.0104)
R2	0.7239	0.6722	0.2342	0.4180	0.4870	0.7414	0.8259	0.4831
F value	7.34***	5.74***	0.85	7.75***	5.70***	8.02***	13.27***	9.91***

(Figures in parentheses indicate the standard errors of the respective coefficients.)
*, ** and *** indicates the level of significance at 10, 5 and 1 per cent. NS = non-significant.

Appendix-IX(C)

Results of Cobb-Douglas production function of pearlmillet in different agro-climatic zones

Particulars	*1980-81*				*2000-01*			
	V	*VI*	*VII*	*Overall*	*V*	*VI*	*VII*	*Overall*
Constant (a)	-0.8433 (1.1641)	-4.5573 (0.8664)	0.4011 (0.4260)	-0.7929 (0.5911)	0.6726 (0.4702)	-0.3165 (0.2231)	-0.4851 (0.8136)	-0.2046 (0.5295)
Human labour in Mandays (X1)	0.8001** (0.3753)	1.4053** (0.7479)	0.1839 (0.1726)	0.8014*** (0.2749)	0.3027** (0.1413)	0.6471*** (0.0989)	0.4852** (0.2741)	0.4702*** (0.0787)
Bullock power in pairdays (X2)	0.1502 (0.9097)	2.2536*** (0.8384)	0.1354 (0.2366)	0.5515*** (0.1486)	0.0959 (0.1706)	0.0697* (0.0410)	-0.2179 (0.3748)	0.6005*** (0.0421)
Seed in Kg. (X3)	-0.0369 (0.6251)	0.4055 (0.4527)	0.0079 (0.1271)	0.4005 (0.3512)	-0.4213 (0.4483)	-0.3279 (0.3937)	0.0305 (1.0614)	-0.1311 (0.2750)
Manures and fertlizers in Rs. (X4)	0.0217 (0.0666)	-0.1257*** (0.0453)	0.0017 (0.0109)	-0.0104 (0.0355)	0.0593** (0.0219)	-0.0085 (0.0179)	0.0146 (0.0314)	0.0066 (0.0141)
Other working capital in Rs. (X5)	-0.0610 (0.1424)	0.1448 (0.1144)	-0.0282 (0.0404)	-0.0620 (0.0906)	-0.0206 (0.1160)	0.1456** (0.0761)	0.2826** (0.1488)	0.1460*** (0.0594)
R2	0.4081	0.8160	0.1662	0.4767	0.7652	0.8799	0.7494	0.7248
F value	1.93	12.42***	0.31	8.63***	9.12***	20.51***	4.78***	25.28***

(Figures in parentheses indicate the standard errors of the respective coefficients.)
*, ** and *** indicates the level of significance at 10, 5 and 1 per cent. NS = non-significant.

Appendix-IX(D)

Results of Cobb-Douglas production function of groundnut in different agro-climatic zones

Particulars	1980-81			2000-01		
	V	*VII*	*Overall*	*V*	*VII*	*Overall*
Constant (a)	-9.6224 (2.7477)	-0.0777 (3.8728)	-0.4253 (2.8437)	0.5930 (0.8895)	1.9651 (2.2236)	-1.2215 (0.9475)
Human labour in Mandays (X1)	0.9430 (0.8117)	2.0664** (0.9986)	0.6971*** (0.2073)	0.4470** (0.2405)	0.7352*** (0.1851)	0.5138*** (0.1333)
Bullock power in pairdays (X2)	1.4900** (0.5964)	-1.0917* (0.7620)	-0.4599* (0.2975)	-0.0218 (0.1780)	0.2188 (0.2481)	0.1076* (0.0676)
Seed in Kg. (X3)	2.3630* (1.5347)	-1.2446 (1.4260)	-0.2069 (0.2332)	-0.3684 (0.3381)	-1.2907 (1.0968)	0.0988 (0.4297)
Manures and fertlizers in Rs. (X4)	0.1380** (0.0675)	0.2189*** (0.0732)	0.2193*** (0.0621)	0.1074*** (0.0279)	0.0851* (0.0524)	0.1763*** (0.0588)
Other working capital in Rs. (X5)	-0.1608** (0.0722)	0.3648*** (0.1050)	0.2304*** (0.0796)	0.0481 (0.8538)	-0.0553 (0.1024)	0.0195 (0.0495)
R2	0.8506	0.8741	0.8357	0.8482	0.6506	0.8094
F value	15.93***	20.83***	35.60***	15.64***	5.21***	28.87***

(Figures in parentheses indicate the standard errors of the respective coefficients.)
*, ** and *** indicates the level of significance at 10, 5 and 1 per cent. NS = non-significant.

Appendix-IX(E)

Results of Cobb-Douglas production function of cotton in different agro-climatic zones

Particulars	*1980-81*			*2000-01*		
	VII	*VIII*	*Overall*	*VII*	*VIII*	*Overall*
Constant (a)	-1.0596 (0.3943)	-0.8676 (0.4855)	-0.9611 (0.3269)	0.3391 (0.9256)	-0.5042 (0.5023)	-0.2386 (0.4921)
Human labour in Mandays (X1)	0.8931** (0.3587)	0.5244 (0.4351)	0.7231*** (0.2851)	0.1778 (0.5028)	0.0668 (0.1822)	0.4985** (0.2222)
Bullock power in pairdays (X2)	-0.0111 (0.5104)	0.3233 (0.4266)	0.2169 (0.3064)	-0.0616 (0.2297)	0.7012*** (0.1916)	0.1996 (0.1791)
Seed in Kg. (X3)	-0.1184 (0.1133)	-0.2453** (0.1139)	-0.1808** (0.0785)	0.2806** (0.1639)	-0.1307 (0.1001)	0.0945 (0.1198)
Manures and fertlizers in Rs. (X4)	0.1149** (0.0538)	0.0413 (0.0332)	0.0701*** (0.0260)	0.0768 (0.0548)	0.0701 (0.0801)	0.0344 (0.0392)
Other working capital in Rs. (X5)	-0.0419 (0.0768)	0.1060 (0.0611)	0.0259 (0.0474)	-0.0534 (0.1278)	0.0873* (0.0618)	0.0520 (0.0716)
R2	0.7264	0.8045	0.6668	0.4379	0.8738	0.4148
F value	6.90***	11.52***	13.20***	2.02	19.39***	4.67***

(Figures in parentheses indicate the standard errors of the respective coefficients.)
*, ** and *** indicates the level of significance at 10, 5 and 1 per cent. NS = non-significant.

Appendix-IX(F)

Results of Cobb-Douglas production function of *rabi* sorghum in different agro-climatic zones

Particulars	*1980-81*			*2000-01*		
	VI	*VII*	*Overall*	*VI*	*VII*	*Overall*
Constant (a)	-0.5212 (1.0659)	0.6419 (0.5946)	-0.469 (0.4265)	-1.6214 (0.6753)	-0.2121 (1.0469)	-0.6942 (0.4592)
Human labour in Mandays (X1)	0.1649 (0.3720)	0.2034 (0.2995)	0.1427 (0.2514)	0.1604 (0.1674)	0.3667*** (0.1288)	0.1565** (0.0814)
Bullock power in pairdays (X2)	2.0787*** (0.5595)	0.5232* (0.3154)	-0.2624* (0.1592)	-0.0008 (0.1520)	-0.0709 (0.1788)	-0.2116** (0.0955)
Seed in Kg. (X3)	-0.6613 (1.0542)	-0.5778** (0.2868)	-1.1004 (0.6508)	1.8338*** (0.4338)	0.0224 (0.8963)	-1.0702*** (0.3451)
Manures and fertlizers in Rs. (X4)	0.0027 (0.0426)	0.1304*** (0.0235)	0.1409*** (0.0256)	0.0847** (0.0414)	-0.0322 (0.0330)	0.1614* (0.0683)
Other working capital in Rs. (X5)	-0.0748 (0.1078)	-0.0236 (0.0635)	0.0689 (0.0604)	0.1009 (0.1114)	0.2257 (0.2230)	-0.0509 (0.0888)
R2	0.6112	0.9016	0.718	0.8022	0.7763	0.7365
F value	4.40***	18.32***	15.27***	11.35***	6.94***	16.76***

(Figures in parentheses indicate the standard errors of the respective coefficients.)

*, ** and *** indicates the level of significance at 10, 5 and 1 per cent. NS = non-significant.

Appendix-IX(G)

Results of Cobb-Douglas production function of wheat in different agro-climatic zones

Particulars	1980-81				2000-01			
	VI	*VII*	*VIII*	*Overall*	*VI*	*VII*	*VIII*	*Overall*
Constant (a)	-1.6942 (1.6347)	-4.8680 (5.2667)	-2.3211 (5.2336)	1.0896 (3.7456)	-0.5086 (1.4712)	0.4334 (1.1075)	-1.4566 (0.4700)	1.1785 (0.5514)
Human labour in Mandays (X1)	1.0244*** (0.3905)	2.5531* (1.5894)	0.6636 (2.4763)	0.3604 (1.1631)	0.1622 (0.1342)	0.1716 (0.1574)	0.2680** (0.1481)	0.3926*** (0.0843)
Bullock power in pairdays (X2)	-0.2265 (0.1844)	0.1709 (0.8140)	-1.6077* (1.0976)	-0.0153 (0.5520)	0.0330 (0.0378)	0.0569 (0.0649)	0.2840** (0.1624)	0.0128 (0.0254)
Seed in Kg. (X3)	0.2107 (0.9343)	0.6619 (2.4316)	1.5908 (2.2843)	0.2184 (0.7703)	0.1541 (0.6773)	-0.3621 (0.5310)	0.2339 (0.2150)	0.1174 (0.2562)
Manures and fertlizers in Rs. (X4)	-0.0685* (0.0504)	-0.0318 (0.1696)	0.1422* (0.0863)	0.1942*** (0.0566)	0.0252 (0.0230)	0.1023*** (0.0255)	0.1570** (0.0740)	0.2300*** (0.0815)
Other working capital in Rs. (X5)	0.3486*** (0.0909)	0.0050 (0.1865)	0.0508 (0.3402)	0.2009** (0.0981)	0.3473** (0.1501)	0.2757* (0.1789)	0.2758*** (0.0832)	0.2103*** (0.0777)
R2	0.7426	0.3478	0.2830	0.3790	0.5719	0.7768	0.8509	0.6026
F value	6.92***	1.27	1.02	5.98***	3.20**	8.35***	14.83***	14.86***

(Figures in parentheses indicate the standard errors of the respective coefficients.)

*, ** and *** indicates the level of significance at 10, 5 and 1 per cent. NS = non-significant.

Appendix-IX(H)

Results of Cobb-Douglas production function of *suru* sugarcane in different agro-climatic zones

Particulars	*1980-81*			*2000-01*		
	VI	*VII*	*Overall*	*VI*	*VII*	*Overall*
Constant (a)	2.5216 (0.6622)	1.3383 (0.6784)	1.0427 (0.5560)	1.4820 (0.6470)	2.5908 (1.1727)	1.1733 (0.5336)
Human labour in Mandays (X1)	0.0318 (0.0774)	0.0726 (0.0611)	0.2106* (0.1355)	-0.2745** (0.1236)	-0.0549 (0.1492)	0.2754** (0.1021)
Bullock power in pairdays (X2)	-0.0025 (0.0251)	0.0445** (0.0188)	0.0741 (0.0442)	0.0201 (0.0298)	0.0102 (0.0324)	0.0912** (0.0489)
Seed in Kg. (X3)	0.0361 (0.1189)	0.3083** (0.1287)	0.1330 (0.1029)	-0.2970** (0.1689)	-0.3310* (0.1931)	0.1015** (0.0471)
Manures and fertlizers in Rs. (X4)	0.0321 (0.0315)	0.0468** (0.0201)	0.0876** (0.0445)	0.0870 (0.0675)	-0.0305 (0.0657)	0.0862 (0.0929)
Other working capital in Rs. (X5)	0.0314 (0.0458)	-0.0027 (0.0255)	0.1449* (0.0748)	-0.0127 (0.0297)	-0.0308 (0.0530)	0.1139* (0.0793)
R2	0.1497	0.7780	0.3801	0.4554	0.5486	0.4542
F value	0.49	4.09***	2.02	2.34**	1.45	2.85**

(Figures in parentheses indicate the standard errors of the respective coefficients.)

*, ** and *** indicates the level of significance at 10, 5 and 1 per cent. NS = non-significant.

Appendix-IX(I)

Results of Cobb-Douglas production function of *ratoon* sugarcane in different agro-climatic zones

Particulars	*1980-81*			*2000-01*		
	VI	*VII*	*Overall*	*VI*	*VII*	*Overall*
Constant (a)	1.7016 (0.5647)	3.1279 (0.2628)	1.9561 (0.3096)	1.7772 (0.6600)	2.1172 (0.3115)	1.9766 (0.2541)
Human labour in Mandays (X1)	0.1633 (0.1215)	0.1023 (0.0909)	0.2146** (0.0916)	0.2787** (0.1306)	0.3356*** (0.0877)	0.2882*** (0.0596)
Bullock power in pairdays (X2)	0.0009 (0.0218)	0.0659*** (0.0233)	0.0171* (0.0126)	0.0072 (0.0279)	0.0092 (0.0208)	0.0157 (0.0141)
Manures and fertlizers in Rs. (X3)	0.0885 (0.091)	0.0293 (0.0560)	0.1143 (0.0938)	0.0521 (0.0887)	0.025 (0.0434)	0.0665** (0.0357)
Other working capital in Rs. (X4)	0.1280 (0.1043)	-0.1888*** (0.0674)	-0.0171 (0.0636)	0.0948 (0.0792)	-0.0364 (0.0538)	-0.0259 (0.0372)
R2	0.5959	0.5959	0.3847	0.3659	0.5818	0.5012
F value	1.28	4.79***	2.97*	1.58	3.82**	7.28***

(Figures in parentheses indicate the standard errors of the respective coefficients.)

*, ** and *** indicates the level of significance at 10, 5 and 1 per cent. NS = non-significant.

Appendix-X

Results of Chow test for old and new period technology functions at overall level

Sr. No.	*Crop*	*Degrees of freedom*	*F- ratio*
1	*Kharif* Sorghum	80	8.30***
2	*Kharif* paddy	113	16.18***
3	Pearlmillet	102	15.46***
4	Groundnut	74	10.62***
5	Cotton	72	5.68***
6	Rabi sorghum	66	7.84***
7	Wheat	104	17.03***
8	Sugarcane *(Suru)*	58	4.47***
9	Sugarcane *(Ratoon)*	63	7.95***

(Note- * indicate the level of significance at one per cent)**

Appendix -XI: Geometric Mean Levels of Inputs and Outputs of Major Crops in Different Agro-Climatic Zones of Maharashtra

Crop	*Zone*	*Human labour (mandays)*		*Bullock power (pairdays)*		*Seed (kgs.)*		*Manure & fertlizers (Rs.)*		*Other working capital (Rs.)*		*Output (Qtls.)*	
Kh. sorghum	VII	99.70	45.95	18.35	4.67	11.72	8.45	770.86	1614.61	386.49	994.11	14.70	26.16
	VIII	50.30	86.95	12.24	5.97	9.24	9.94	52.09	1166.09	36.86	1483.12	6.63	17.11
	Overall	71.38	91.20	15.06	5.26	10.43	9.15	206.75	1377.35	122.67	1208.61	9.97	21.26
Kh. paddy	II	119.17	45.25	11.65	8.23	44.90	91.82	507.70	1525.67	332.95	379.35	8.39	20.13
	V	189.86	81.60	12.81	2.49	46.63	94.36	1715.81	538.32	922.92	380.18	11.90	22.21
	IX	115.46	118.45	10.82	2.76	44.76	102.07	271.73	1171.36	60.69	648.10	13.90	22.60
	Overall	137.72	76.57	11.73	3.79	45.42	96.06	618.58	976.93	265.15	455.21	11.15	21.64
Pearlmillet	V	33.15	73.67	13.01	15.81	3.55	3.81	5.12	165.33	103.29	554.93	2.72	15.25
	VI	43.58	53.97	12.77	9.12	3.52	3.99	22.91	168.80	123.28	582.65	3.81	11.43
	VII	46.58	58.36	13.44	8.26	3.67	3.91	36.46	197.11	238.93	436.65	6.44	9.33
	Overall	40.07	61.80	13.03	10.90	3.57	3.90	14.83	174.38	137.10	530.83	3.85	12.06
Groundnut	V	100.35	130.12	18.93	2.06	112.04	101.11	1748.56	2854.26	751.84	1141.19	11.47	20.49
	VII	86.25	82.67	14.37	1.59	117.98	99.27	499.84	834.23	372.41	590.38	3.37	8.43
	Overall	92.87	103.71	16.43	1.81	115.04	100.19	920.72	1543.09	524.65	820.81	6.13	13.14
Cotton	VII	53.15	102.34	11.79	13.04	4.89	3.03	310.44	985.48	83.87	687.51	3.52	6.94
	VIII	53.20	74.21	13.84	13.14	2.96	5.70	165.26	1813.06	245.51	384.26	3.56	5.76
	Overall	53.18	86.79	12.80	13.09	3.78	4.19	303.82	1347.18	145.49	510.17	3.54	6.31

(Contd.)

(Contd.)

Crop	Zone	Human labour		Bullock power (mandays)		Seed (kgs.) (pairdays)		Manure & fertlizers (Rs.)		Other working capital (Rs.)		Output (Qtls.)	
Rabi sorghum	VI	24.85	64.92	8.01	5.40	10.70	11.02	32.05	125.77	145.04	714.24	6.24	11.11
	VII	26.88	70.05	10.51	9.07	11.35	11.60	51.29	7.97	93.27	463.32	9.23	9.84
	Overall	25.74	67.15	9.04	6.80	10.99	11.27	39.51	36.91	119.18	589.26	7.43	10.53
Wheat	VI	63.35	95.86	15.75	5.99	108.33	104.80	2822.93	349.19	4094.38	2150.52	14.16	23.51
	VII	66.85	80.88	16.67	4.01	107.94	102.45	1831.10	466.02	3542.32	1785.30	18.94	17.25
	VIII	64.90	79.15	15.44	14.93	109.58	106.83	301.26	1647.86	2610.22	1488.95	4.14	17.39
	Overall	62.02	84.87	15.93	7.20	108.63	104.72	1131.04	655.95	3342.41	1782.10	10.19	19.04
Sugarcane *(Suru)*	VI	429.02	301.40	19.35	9.74	7689.15	7338.28	8695.93	8550.94	11577.61	7820.84	942.16	937.74
	VII	457.85	254.08	23.64	7.98	7635.48	7408.93	6280.11	9033.04	15180.19	9244.11	853.53	907.06
	Overall	436.90	282.70	21.95	9.04	7681.61	7364.70	7895.32	8788.74	11136.74	8522.88	907.89	926.12
S. cane *(Ratton)*	VI	238.71	252.84	5.09	2.50	0.00	0.00	5532.15	6192.61	10330.83	4682.92	763.15	823.66
	VII	297.68	217.58	5.29	5.17	0.00	0.00	4387.87	5396.14	7945.55	4667.94	711.17	784.01
	Overall	261.87	234.54	5.21	3.60	0.00	0.00	5020.57	5640.02	9264.41	4672.53	737.44	803.59

Bibliography

Agarwal R. C. 1983. Role of farmer's education in increasing agricultural production in the Himalayan region of Nepal and India. *Paper presented at the fifth International Farm Management Congress*. Famam Royal, London.

Alauddin M. and C. A. Tisdell. 1988. Impact of new agricultural technology on the instability of foodgrain production and yield: Data analysis for Bangladesh and its districts. *Journal of Development Economics*. 29 (2) : 199-227.

Ali M. A.; K. Brahmachari; R. K. Bandyopadhyay; D. S. Gupta and T. K. Nath. 2002. An analysis of socio-economic constraints to the adoption of modern agricultural technology in Dakshin Dinajpur District of West Bengal: A case study. *Journal of Inter Academicia*. 6 (3) : 340-347.

Al-Karablieh E. 1995. An assessment of the impact of agricultural technology on output in the rainfed farming areas in Jordan. *Wissenschaftsverlag Vauk Kiel KG*; Kiel; Germany. v + 269. (Quotation from *Agricultural Economics Abstracts, 1996*)

Alshi M. R.; P. Kumar and V. C. Mathur. 1983. Technological change and factor share in cotton production : A case study of Akola cotton farms. *Indian Journal of Agricultural Economics*. 38 (3) : 407-415.

Badal P. S. and R. P. Singh. 2001. Technological change in maize production. *Indian Journal of Agricultural Economics*. 56 (2) : 211-219.

Baliyan S. P.; T. S. Bhogal and S. Pal. 1998. Resource use efficiency in planted crop of sugarcane in western Uttar Pradesh. *Indian Sugar*. 48 (5) : 329-336.

Bhalerao P. D.; P. N. Jadhav and G. R. Fulzele. 1993. Productivity and economics of sunflower-pigeon pea intercropping under rainfed conditions. *Panjabrao Krishi Vidyapeeth Research Journal.* 17 (1):23-36.

Birari K. S. 1999. Resource use structure, resource productivities and allocation efficiency on farms of Western Maharashtra. *Ph. D. Thesis submitted to M. P .K. V. Rahuri (MS).*

Bisaliah S. 1977. Decomposition analysis of output change under new production technology in wheat farming : Some implications to returns on research investment. *Indian Journal of Agricultural Economics.* 32 (3): 193-201.

Bisaliah V. and B. M. Sharma. 1989. HYV adoption, productivity gaps and production adjustment mechanism in flood-prone area of Uttar Pradesh - A synthesis. *Research paper No. 5, Institute of Rsarali Management, Anand, Gujrath.*

Bogahawatte C. 1992. Modern technology and variability of paddy production in Sri Lanka. *Indian Journal of Agricultural Economics.* 47 (1):90-96.

Bonny B. P. and K. Vijayaragavan. 2001. Adoption of sustainable agricultural practices by traditional rice growers. *Journal of Tropical Agriculture.* 39 (2): 151-156.

Borah K. C. 1983. Factor shares in traditional farming in Assam -A case study in Majuli - A river island. *Indian Journal of Agricultural Economics.* 38 (3): 438.

Brahmachari K.; M. A. Ali, and T. K. Nath. 2003. Socio-technical analysis of the extent of adoption of modern agricultural technology in Dakshin Dinajpur District of West Bengal. *Journal of Environment and Ecology.* 21 (1): 152-155.

Cauvery R. 1991. Groundnut production in Tamil Nadu: a decomposition analysis. *Agricultural Situation in India.* 46 (5) : 321-324.

Caviglia J. 1992. Economic benefits in sustainable technological proposals. *Horizonte Agro Economico.* 1(2) : 40-47. (Quotation from *Agricultural Economic Abstract, 1993*)

Chandra N. and R. P. Singh. 1992. Determinants and impact of new technology adoption in tribal agriculture in Bihar. *Indian Journal of Agricultural Economics.* 47 (3): 397-403.

Chow G. C. 1960. Test of equality between sets of coefficients in two linear regressions. *Econometrica.* 28 (2) : 591-605.

Dantwala M. L. 1970. From stagnation of growth.*Indian Economic Journal.* 18 (2) : 165-192.

Despande R. S. and A. Chandrasekhar. 1983. Growth and supply response of slow growth crop. *Indian Journal of Agricultural Economics.* 47 (3) : 393-397.

Dhindsa K. K. and A. Sharma. 1996. Growth of agricultural production and productivity in Punjab 1970-73 through 1990-93: A component analysis. *Anvesak.* 26 (1) : 51-61.

Falcon W. P. 1970. The green revolution generation of problems. *American Journal of Agricultural Economics.* 52 (5) : 698-709.

Frankel F. R. 1971. India's green revolution : Economic gains and political costs. *Oxford University Press,* Mumbai.

Gaddi G. M.; S. M. Mundinamani and S. A. Patil. 2002. Yield gaps, constraints and potential in cotton production in North Karnataka— An econometric analysis. *Indian Journal of Agricultural Economics.* 57 (4) : 722-734.

Gaikwad D. M. 2001. A Study on impact of the selected dryland technologies on farm productivity in Solapur district. *M. Sc. (Agri.) Thesis submitted to M.P.K.V. Rahuri (MS).*

Gomez A. A; D. E. S. Kelly; J. K. Syers; K. J. Coughlan; J. W. Doran and A. J. Jones. 1996. Measuring sustainability of agricultural systems at the farm level. *'Methods for assessing soil quality'. Book chapter,* 49 (2) : 401-410.

Goswami S. N.; N. C. Khandare; T. N. Hajare and T. K. Sen. 2001. Growth trend of principal crops of Maharashtra: an analytical approach. *Journal of Maharashtra Agricultural University.* 26 (1) : 90-93.

Government of India. 2004. Agricultural statistics. *Ministry of Agriculture, Government of India.* pp : 56-59.

Government of Maharashtra. 2002. *Economic Survey of Maharashtra, 2001-02.*

Gupta D. N. 1998. Need for effective policy of sustainable technology for development of rural areas. *Journal of Rural Development, Hyderabad.* 17 (3) : 511-527

Haffis S.; P. N. Laxmi, R. V. V. S. G. K. Raju, Y. U. R. Reddy and U. M. V. Rao. 1990. Constraints analysis in production of dryland crops. *Indian Journal of Dryland Agricultural Research and Development.* 5 (1 - 2) : 72-77.

Heady E. O. and J. L. Dillon. 1961. Agricultural production functions. *Iowa State University Press, Ames, Iowa, USA* : 667.

Hernandez C. M., J Crossa and A. Castillo. 1993. The area under the function : An index for selecting desirable genotypes. *Theoretical and Applied Genetics.* 87 : 409-415.

Hiremath B. N.; R. V. R. Kesi; M. Flury and U. Geiser. 2002. Technology adoption in semi-arid Gujarat: The importance of livelihood security. *Local environmental management in a North-South perspective* : 203-212.

Johnston J. 1972. Econometric methods. McGraw-Hill Book Company, (Second edition), New York.

Kale P. V. and D. L. Sale. 2000. Resource use efficiency on sugarcane farms in western Maharashtra. *Indian Sugar.* 49 (11) : 909-912.

Kale P. V.; B. S. Kakad; B. N. Pawar and D. L. Sale. 1999. Changes in resource use and productivity levels of sugarcane in Western Maharashtra. *Indian Sugar.* 49 (3) : 193-198.

Kanwar J. S. 2000. Prospective and policies for land use planning. *Journal of Indian Society of Soil Science.* 48 (4) : 636-639.

Kapase, P. M. 1992. Differentials in resource use structure and productivities of principal Crops in different agro-climatic zones of Maharashtra. *Ph. D. Thesis submitted to M.P.K.V. Rahuri (MS).*

Karagiannis G.; S. Katranidis and K. Velentzas. 1996. Decomposition analysis of factor cost shares: the case of Greek agriculture. *Journal of Agricultural and Applied Economics.* 28 (2): 369-379.

Katyal V.; K. S. Gangwar and B. Gangwar. 1999. Influence of input use and management practices on sustainability and economics of rice cultivation. *Journal of the Andaman Science Association.* 15 (1) : 56-59.

Khan B. A. and S. A. Alam. 1998. Impact of technological change on size productivity relationship and resource use efficiency in Kashmir agriculture. *Agricultural Situation in India.* 43 (4): 301-307.

Kiresur V.; P. S. Reddy, M. V. R. Prasad and B. N. Reddy. 1995. Economic sustainability of improved oilseed production technologies under real farm situations. *Bulletin, Directorate of Oilseed Research, Hyderabad.* 32 p.

Kiresur V.; R. Balakrishnan and M. V. R. Prasad. 1996. A model for estimation of economic sustainability of improved oilseeds crop production technologies. *Indian Journal of Agricultural Economics.* 51 (3) : 328–341.

Kiresur V.; S. V. Ramana Rao and D. M. Hegade. 2001. Improved technologies in oilseeds production—An assessment of their economic potentials in India. *Agricultural Economics Research Review.* 14 (2) : 95-108.

Kiresur V.; R. K. Pandey and Mruthyunjaya. 1995. Technological change in sorghum production: an econometric study of Dharwad farmers in Karnataka. *Indian Journal of Agricultural Economics.* 50 (2) : 185-192.

Klein L. R. 1953. An Introduction to Econometrics. Book. Harper and Row, New York.

Koutsoyiannis A. 1977. Theory of Econometrics. The Macmillan Press Ltd. London. Second Edition. 675 p.

Krishna Rao. 2000. Yield gaps and constraints for low yields of selected crops in Anantpur district of Andhra Pradesh. *International Journal of Tropical Agriculture.* 20 (1-4): 107-113.

Lalwani M. 1990. Human labour absorption in dairying-Evidence from Karnal village of Haryana. *Indian Journal of Agricultural Economics.* 45 (2) : 150-158.

Malik S. K. and R. P. Singh. 1999. Resource use efficiency and productivity in sugarcane production in western U.P. *Cooperative Sugar.* 30 (8) : 774-776.

Mitra S. and S. Mitra. 2002. Resource use efficiency of Barga farms: an impediment to agricultural growth. *Economic Affairs Calcutta.* 47 (2) : 78-87.

Nagure D. V.; R. D. Shelake and S. P. Kalyanakar. 2002. Cost of cultivation and gap analysis in adoption of recommended technology for important crops in Sarola village of Parabhani district of Maharashtra. *Maharashtra Journal of Agricultural Economics.* 11 (1) : 25.

Pandey U. K. and M. A. Muralidharan. 1977. An application of discriminant function in agricultural finance. *Indian Journal of Agricultural Economics.* 32 (2) : 41-51.

Pattanaik B. K. 1994. Adoption of modern technology and agricultural productivity—An evaluation study. *Indian Journal of Agricultural Economics.* 49 (3) : 363-364.

Paroda R. S. 1998. Meeting new challenges in food production. *Yojana.* 42 (8).

Parthasarathy D. 2002. Measuring or defining sustainability? Impact assessment of an agricultural technology. *Impact Assessment and Project Appraisal. Report of Govt. of India.* 20 (4) : 293-298.

Patel R. K., A. S. Sirohi and B. M. Sharma. 1968. Productivity and allocation of resources in the production of hybrid bajra in Delhi territory. *Indian Journal of Agricultural Economics.* 23 (4) : 146.

Patil S. M.; L. B. Kunnal and G. M. Gaddi. 1997. Resource use efficiency in groundnut production. *Karnataka Journal of Agricultural Sciences.* 10 (3) : 807-810.

Patil S. S. and L. B. Kunnal. 1988. Cost-return structure of jowar production. *Farming Systems.* 4 (1-2) : 34-37.

Pawar B. N.; H. R. Shinde and D. L. Sale. 2000. Resource productivity and sustainability of sugarcane in western Maharashtra. *Indian-Sugar.* 50 (3) : 147-151.

Pawar J. R. 1978. Changes in resource use structure, resource productivities and allocation efficiency on farms in Maharashtra. *Ph.D. Thesis submitted to M.P.K.V. Rahuri (MS).*

Raghuwanshi R. S.; P. K. Awasthi and P. Sharma. 1999. Resource use efficiency in wheat cultivation. *Indian Journal of Agricultural Research.* 33 (1) : 67-71.

Rahman S. and S. Rahman. 1999. Technological change and food production sustainability in Bangladesh agriculture. *Bangladesh Journal of Agricultural Economics.* 22 (2) : 1-17.

Rajesh R. 2001. Economic evaluation of dry farming technologies: A study in southern districts of Tamil Nadu. *Indian Journal of Agricultural Economics.* 56 (2) : 220-229.

Rao C. K. 1984. A note on new agricultural technology and farm employment. *Indian Journal of Economics.* 65 (2) : 537-541.

Raut R. C. 1997. Study on the impact of new agricultural technology on employment and income in Western Maharashtra. *Ph.D. Thesis submitted to M.P.K.V. Rahuri (MS).*

Ray G. L.; P. Chatterjee and S. N. Banerjee. 1995. Technological gap and constraints in agricultural technology transfer. *Naya Prakash,* Calcutta; Indias. : 94 p.

Reddy N. R. V. 1995. Technological parameters in paddy production function: A case study of Cuddapah district of Andhra Pradesh. *Indian Journal of Economics.* 75 (4) : 497-503.

Reddy P. P. 2002. Farm costs, income, factor shares and price-cost relationship: an empirical analysis of paddy in Andhra Pradesh. *Journal of Research, ANGRAU.* Hyderabad. 30 (1) : 44-55.

Reddy R.; G. V. T. Raju and A. Janaiah. 1997. The comparative economics of cotton cultivation: A case study in Guntur district of Andhra Pradesh. *Journal of Research, ANGRAU,* Hyderabad. 25 (1) : 46-50.

Reddy V. R. 1994. New agricultural technology, increasing costs and relative shares of labour and other factors: A spatio-temporal analysis of Andhra Pradesh. *Indian Journal of Labour Economics.* 37 (2) : 215-227.

Reddy Y. V. R.; J. C. Katyal; H. Shaikh; S. K. Das and R. P. Singh. 1995. Economics of recommended dryland technology in India. Book chapter. *Sustainable Development of Dryland Agriculture in India.* : 511-522.

Saleth, R. M. and M. S. Swaminathan. 1992. Sustainable livelihood index- A litmus test for sustainable development. *Discussion paper. M. S. Swaminathan Research Foundation, Madras.*

Samaratunga P. A. 1991. Changes of rice production stability with the spread of new technology and its implications: The case of Sri Lanka. *Quarterly Journal of International, Agriculture.* 30 (4) : 334-348.

Sani R.M.; S. Kushwaha; G. Abdullahi and C. Sen. 2002. Economic analysis of small-scale sugarcane (*Saccharum spp.*) production in Bauchi State, Nigeria. *Economic Affairs Calcutta.* 47 (1) : 37-41.

Santha A. M. 1993. A comparative analysis of cost and returns of paddy cultivation for different seasons in Trichur, Kerala. *Madras Agricultural Journal.* 80 (2) : 41-44.

Satyanarayan S.; S. M. Kurmvanshi and S. Soni. 1999. Technological status (adoption pattern) of soybean cultivation in district Sagar of Madhya Pradesh. *Crop Research Hisar*. 18 (1) : 150-154.

Schultz T. W. 1964. Transforming traditional agriculture. Book. Yale University Press, Yale, New Haven.

Sen B. 1972. Opportunities in the green revolution. M. L. Dantwala (Ed.), Agricultural development in developing countries: Comparative experience. *Indian Society of Agricultural Economics,* Mumbai, India. : 262-280.

Shaik H., P. N. Laxmi, R. V. S. J. K. Raju, Y.V. R. Reddy and U. M. B. Rao. 1990. Constraint analysis in production of dryland crops. *Indian Journal of Dryland Agricultural Research and Development.* 5 (1) : 72-77.

Shaikh H.; Y. V. R. Reddy and K. L. Sharma. 2001. Resource use efficiency and returns to scale in production of groundnut (*Arachis hypogaea*) in Andhra Pradesh. *Indian Journal of Dryland Agricultural Research and Development.* 16 (1) :1-8.

Sharma G. C. and B. Gangwar. 2000. Economic analysis of sugarcane production in Khadar area of Western U.P. *Bhartiya Krishi Anusandhan Patrika.* 15 (1-2) : 69-75.

Sharma H. R.; D. S. Thakur and K. Singh. 1987. Economic analysis of tribal agriculture of Himachal Pradesh. *Agricultural Situation in India.* 152 (1) : 35-42.

Sharma J. L.; and J. Singh. 1990. Resource use productivity in Punjab. *Agricultural Economics Affairs*. 35 : 9-14.

Sharma K. P. and D. C. Pant. 1994. Impact of agricultural technology on farm economy. *Indian Journal of Agricultural Economics.* 49 (3) : 373-374.

Shukla N.D.; B. Gangwar; P. P. Mishra; V. Katyal and G. C. Sharma. 1999. Resource use and productivity of sugarcane in Meerut district of Uttar Pradesh. *Journal of the Andaman Science Association.* 15 (1) : 39-44.

Sidhu R. S. and S. S. Grewal. 1990. Factors affecting demand for human labour in Punjab Agriculture. *Indian Journal of Agricultural Economics.* 45 (2) : 125-133.

Singh B. and B. Singh. 1997. Adoption gap in improved agricultural technology in Uttar Pradesh hills. *Indian Journal of Hill Farming* 10 (1-2) :19-22.

Singh I. P. and S. S. Grewal. 1994. Economic efficiency in paddy production in Punjab. *Bihar Journal of Agricultural Marketing.* 2 (1) : 55-64.

Singh K. 1998. Sustainable development: some reflections. *Presidential address in 58th annual conference of Indian Society of Agricultural Economics*, held at MPKV, Rahuri (MS) on December 21-23, 1998.

Singh K. V.; S. Kumar S. and R. N. Malik. 1993. Resource use efficiency of sugarcane farms (a micro-level study of western Uttar Pradesh). *Cooperative Sugar*. 25 (1-2) : 43-45.

Sinha R.P. and B. M. Sharma. 1986. Prospects of increasing farm incomes on tubewell bullock operated farms through new technology and credit. *Agricultural Situation in India.* 40 (3) : 139-142.

Sisodiya A. S.; T. B. Singh and S. B. Nahatkar. 1999. Temporal analysis of decomposition of pigeon-pea production in different agro-climatic regions of Madhya Pradesh. *Bhartiya Krishi Anusandhan Patrika* 14 (1-2) : 15-22.

Sodhiya H. C. 1986. Impact of science and technology on agriculture in Sagar division, Madhya Pradesh. *Indian Journal of Agricultural Economics.* 41 (4): 501.

Sohani D. K. and R. D. Khandarkar. 1970. Pattern of income distribution, saving and expenditure in rural India. *Indian Journal of Agricultural Economics*. 25 (3) : 132-133.

Soni S. N.; S. K. Khatik; K. C. Shukla and O. P. Singh. 1997. Improvement in crop yield by adopting new agricultural technology under lab to land programme in District Sagar (M.P.). *Journal of Soils and Crops.* 7 (1) : 46-49.

Solow R. M. 1957. Technical change and the production function. *Review of Economics and Statistics.* 11 (1).

Suchitra G. 1995. Resource use and productivity of turmeric crop in Naziabad district of Andhra Pradesh. *Agricultural Situation in India*. 49 (10) : 727-730.

Sunandini; H. Shaikh; P. B. Parthasarathi and Y. V. R. Reddy. 1993. Resource productivity and resource use efficiency on paddy farms of Andhra Pradesh. *Agricultural Situation in India*. 47 (11) : 835-840.

Swaminathan M. S. 1991. The road to sustainable Agriculture. *Monogram No. 4,* M. S. Swaminathan Research Foundation, Madras.

Swaminathan M. S. 2000. Towards an evergreen revolution on agriculture. *NFI Bulletin*. 21 (2) : 1-4.

Tiwari R. and D. P. Dheer. 1992. Sugarcane and its problems, cost of cultivation of plant and ratoon sugarcane - A micro study of Bihar state. *Indian Sugar*. 42 (2) : 107-110.

Tiwari R. and D. P. Dheer. 1994. Resource use efficiency of ratoon: Sugarcane farms in north Bihar. *Indian Sugar*. 44 (2) : 109-111.

Umesh K. B. and S. Bisaliah. 1986. Productivity differentials between small and large farms—An economic study. *Asian Economic Review*. 28 (3) : 17-33.

Upender M. 1992. Production elasticity and returns to scale on different farms size groups—A sample study. *Indian Journal of Agricultural Economics*. 47 (3) : 492.

Upendra K.; U. K. Singh, R. P. Sinha and U Kumar. 1998. Resource use efficiency in rice cultivation under different eco-systems – A case study of Pusa block of Samastipur district (Bihar). *Journal of Applied Biology*. 8 (2) : 78-81.

Upendra K.; U. K. Singh, R. P. Sinha and U Kumar. 1999. Economics of rice cultivation under different eco-systems – A case study of Pusa block of Samastipur district (Bihar). *Journal of Applied Biology*. 9 (1) : 97-98.

Vasishta A. K.; C. Singh; G. S. Dhaliwal; R. Arora; N. S. Randhawa and A. K. Dhawan. 1998. Growth in agricultural technology and its impact on environment. *Ecological agriculture and sustainable development. Volume-II.* Proceedings of International Conference on *'Ecological Agriculture: towards Sustainable Development'*, held in Chandigarh, India, 15-17 November, 1998. : 512-520.

Velrasu P.; P. Singh and P. Singh. 1999. Fertilizer use pattern and its impact on crop productivity: a case study of Erode district in Tamil Nadu. *Fertilizer News*. 44 (10) : 29-31.

Wawire N. W. O.; D. W. Norman and M. R. Langemeier. 1996. Rainfed rice production in Western Kenya - Productivity and profitability. *Research Report, Department of Agricultural Economics, Kansas State University, USA*. 22 - 23: 30.

Yadav R. N. and A. C. Gangwar. 1983. Factors affecting adoption of new rice technology in Durbhanga district of Bihar— A discriminant function approach. *Agricultural Situation in India*. 42 (9) : 811-814.

Yerriswamy J. and L. B. Hugar. 2001. Resource use efficiency in paddy production under selected agricultural systems in Tungabhadra Project area, Karnataka. *Karnataka Journal of Agricultural Science.* 14 (1) : 86-89.

Index